Springer Textbooks in Earth Sciences, Geography and Environment

The Springer Textbooks series publishes a broad portfolio of textbooks on Earth Sciences, Geography and Environmental Science. Springer textbooks provide comprehensive introductions as well as in-depth knowledge for advanced studies. A clear, reader-friendly layout and features such as end-of-chapter summaries, work examples, exercises, and glossaries help the reader to access the subject. Springer textbooks are essential for students, researchers and applied scientists.

More information about this series at https://link.springer.com/bookseries/15201

Thomas Flüeler

Governance of Radioactive Waste, Special Waste and Carbon Storage

Literacy in Dealing with Long-term Controversial Sociotechnical Issues

 Springer

Thomas Flüeler
Department of Environmental Systems Science
Institute for Environmental Decisions
ETH Zürich
Zürich, Switzerland

Dr. sc. nat. ETH
Umweltrecherchen & -gutachten
Esslingen, Switzerland

ISSN 2510-1307   ISSN 2510-1315   (electronic)
Springer Textbooks in Earth Sciences, Geography and Environment
ISBN 978-3-031-03904-1   ISBN 978-3-031-03902-7   (eBook)
https://doi.org/10.1007/978-3-031-03902-7

Preface

The necessity of being experienced
introduces into knowledge an irrational element,
which cannot be logically justified.

Ludwik Fleck, 1935[1]

Mission impossible. Tampering with Nature. Armchair tangle by a technocrat. Social engineering. Reactions by readers I can imagine. But the fact is that different—toxic—wastes or unwanted substances exist; we produced them and we have to appropriately handle them. Handle them so that succeeding generations—of technicians, engineers, scientists, politicians, affected groups, regions, etc.—will be able, competent and willing to take over. At any rate, they will have to—because the wastes, objectively, are so enduring. We have to convince them that we have done the best possible, so they will not just get a can of worms (or worse).

In times of Coronavirus, it is painful to learn and adapt, but the pandemic challenge also demonstrates some courage of politicians worldwide taking strict measures to curb the virus, at least mitigate its impact. Imagine that the politicians' clampdown were as drastic in tackling climate change and/or waste issues! Within a few years, globally, we would be on a sustainable transformation path. Are our governments, the EU, China, the US, the UN, enterprises, are we all courageous enough to take this step? We should. It lies in our hands—our societies' success in credibly addressing intragenerational (today's) issues might convince future (tomorrow's) generations to be willing to carry on the programmes when needed. After all, the way we deal with waste reflects the way—success or failure—we treat Earth, ourselves and how we take care of our children and grandchildren.

Zürich, Switzerland

Thomas Flüeler

---

[1]Fleck L (1979) Genesis and development of a scientific fact. Trenn TJ, Merton RK (eds), transl. Bradley F, Trenn TJ. University of Chicago Press, Chicago, IL, 95f.

Orig.: Fleck L (1935, 1980) Entstehung und Entwicklung einer wissenschaftlichen Tatsache. Einführung in die Lehre vom Denkstil und Denkkollektiv. Schäfer L, Schnelle T (eds). Suhrkamp Taschenbuch Wissenschaft, Frankfurt aM, 125:

"Die Notwendigkeit der Erfahrenheit bringt ein irrationales, logisch nicht legitimierbares Element in das Wissen."

Acknowledgements

Through time, I have been grateful for discussions, debates and ideas I was able to share with a multitude of people. Particularly, however, I would like to emphasise the exchange with the reviewers, namely Dr. Paul Bossart, former Director of the Underground Research Laboratory Mont Terri; Prof. Paul Brunner, Waste and Resource Management, Technical University Vienna; Prof. Alan G. Green, Engineering and Environmental Geophysics, ETH Zürich; Dr. Peter Hocke, Head of the research group "Radioactive Waste Management as a Socio-Technical Project", ITAS, Institute for Technology Assessment and Systems Analysis, Karlsruhe Institute of Technology; Dr. Daniel Metlay, Senior Fellow, B. John Garrick Institute of Risk Sciences, University of California, Los Angeles, former Senior Staff of US Nuclear Waste Technical Review Board; Dr. Claudio Pescatore, former principal administrator at NEA/OECD, now with Linnaeus University, Sweden, at UNESCO Chair of Heritage Futures; Dr. Felix Altorfer, Swiss Federal Nuclear Safety Inspectorate ENSI; Prof. Didier Sornette, ETH; and Prof. Helmut Weissert, ETH. I am equally indebted to Doris Bleier and Jayanthi Krishnamoorthi of Springer Nature Publishers.

About This Book

The necessity of being experienced
introduces into knowledge an irrational element,
which cannot be logically justified.
Introduction to a field of knowledge
is a kind of initiation that is performed by others. It opens the door.
But it is individual experience,
which can only be acquired personally,
that yields the capacity for active and independent cognition.

Ludwik Fleck, 1935[2]

Deep geological repositories of nuclear waste, long-term landfills of special waste and the substantive storage of carbon dioxide to relieve the world's climate from excessive system change are intricate and contentious policy fields with an impact of decades to hundreds of thousands of years. The present work aims to explore how society and technology can set up and implement sustainable ways to cope with the issues.

Didactically, the aim is to formulate requisites for governance to be followed up, developed and empirically tested in subsequent courses, lectures and research at universities, colleges, public administrations and with senior policy staff of government and companies, and eventually in other fields of application.

The approach provides all necessary background information to be sensibly used. Readers need not be specialists in a particular field; therefore, the present guide is appropriate for various audiences—practitioners, researchers, policymakers, research policymakers, (technical) graduates and (non-technical) undergraduates. Respective knowledge and skills for "handing over the torch" are so far neither a part of academic curricula nor of competences in industrial companies or government services. We have to become long-term literates.

The marketing strategy to make the book accessible on a chapter-by-chapter basis implies some redundancies. First readers may limit themselves to key references; further literature structured by theme enables a focused in-depth study.

---

[2]Fleck L (1979) Genesis and development of a scientific fact, 95f.
　Orig.: Fleck L (1935, 1980) Entstehung und Entwicklung einer wissenschaftlichen Tatsache. Einführung in die Lehre vom Denkstil und Denkkollektiv, 125f:
　"Die Notwendigkeit der Erfahrenheit bringt ein irrationales, logisch nicht legitimierbares Element in das Wissen. Die Einführung, eine Art Weihe, die andere erteilen, eröffnet den Eingang in ein Wissen,—Erfahrung, immer nur persönlich erlangbar, befähigt erst zum tätigen, selbständigen Erkennen."

Summary

Deep geological repositories of nuclear waste, long-term landfills of special waste and the substantive storage of carbon dioxide to relieve the world's climate from excessive system change are intricate and contentious policy fields with an impact of decades to hundreds of thousands of years. This work aims to explore how society and technology can set up and implement sustainable ways—in the long run—to cope with the issues. They arc all long term safety issues and require long-term institutional involvement of the technoscientific community, waste producers, public administrators, NGOs and the public. The demonstration of long-term safety is challenging, and monitoring may contribute to substantiate evidence, support decision making and legitimise the programme. What, where and when to monitor is determined by its goal setting: it may be operational, confirmatory (in the near field) or environmental (far field). As it is "difficult to make predictions, especially about the future",[3] this contribution, too, does not pretend to present the silver bullet, but Strategic Monitoring as proposed and developed contributes to process, implementation or policy and institutional surveillance to sustain a once launched programme. It not only addresses the controversial long-lasting "problem" (of nuclear, toxic or CO_2 waste) but also investigates some ways to approach for "solutions" or solution spaces, not only technical but also institutional and personal, and this is for the long term. It includes the tailored transfer of knowledge, concept and system understanding, experience and documentation to specific audiences mentioned above. It is an integrative tool of targeted, yet adaptive, management and may be applicable to other long-term sociotechnical fields.

---

[3]Famous saying of Danish origin, proverb or Niels Bohr, also attributed to Mark Twain, Samuel Goldwyn or Nostradamus (https://quoteinvestigator.com/2013/10/20/no-predict/, accessed 27 January 2023).

Contents

Graphical Contents

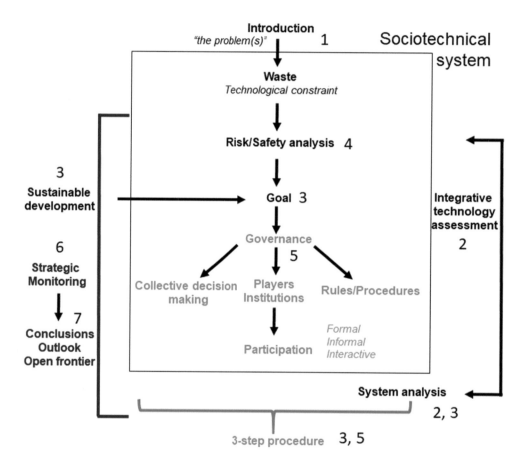

About the Author

The necessity of being experienced
introduces into knowledge an irrational element,
which cannot be logically justified.
Introduction to a field of knowledge
is a kind of initiation that is performed by others. It opens the door.
But it is individual experience,
which can only be acquired personally,
that yields the capacity for active and independent cognition.
The inexperienced individual merely learns but does not discern.

Ludwik Fleck, 1935[4]

Thomas Flüeler was trained in earth and life sciences, risk analysis and science-technology studies. In his professional life, he has gained insights particularly into the issues covered in this book through distinctly different perspectives (see Annex, "key references", chronologically listed, denote the major own argumentative and empirical bases in each chapter):

- Science journalism (free lancer, editor, 1983–1986/1990);
- Environmental NGOs (director of Swiss Energy Foundation, 1986–1990);
- Conflict resolution issues and techniques: mandate by NGOs in a federal mediation attempt "conflict-solving group radioactive waste" (1991–1992);
- Consultancy (independent advisor, from 1991);
- Scientific, technical and policy advice to Government (Swiss Federal Nuclear Safety Commission, chair of Standing committee on radiation protection and waste disposal, 1992–2004, 2001–2004);
- Expertise in technical support organisations: International Atomic Energy Agency, IAEA; Nuclear Energy Agency, NEA; German Federal Office for Radiation Protection (BfS), Slovak Center for Nuclear Safety, German National Citizens' Oversight Committee, etc. (from 2001);
- Advice to Cantonal Government (Cantonal Technical Group Wellenberg, 2000–2002);
- Academia: research and teaching (ETH Zürich, from 1996);

---

[4]Fleck L (1979) Genesis and development of a scientific fact, 95f.

Orig.: Fleck L (1935, 1980) Entstehung und Entwicklung einer wissenschaftlichen Tatsache. Einführung in die Lehre vom Denkstil und Denkkollektiv, 125f:

"Die Notwendigkeit der Erfahrenheit bringt ein irrationales, logisch nicht legitimierbares Element in das Wissen. Die Einführung, eine Art Weihe, die andere erteilen, eröffnet den Eingang in ein Wissen,—Erfahrung, immer nur persönlich erlangbar, befähigt erst zum tätigen, selbständigen Erkennen. Der Unerfahrene lernt nur, er erkennt nicht."

- Public administration, civil service (Directorate of Public Works of the Canton of Zürich, Nuclear Technology Unit Head, from 2009 to October 2022);
- Stakeholding and technical expertise (partial state, i.e., canton) in the current Swiss site selection process for nuclear repositories (from 2009 to October 2022);
- Support for decision making regarding emergency and large-scale events and situations: Federal Civil Protection Crisis Management Board (member on behalf of the Conference of the Cantonal Energy Ministers) (from 2011 to October 2022);
- Oversight of Underground Research Laboratory (Commission de suivi de la République et Canton du Jura: supervision of URL and advice to Government regarding the URL Mont-Terri, Switzerland, from 2012).

Abbreviations

AkEnd	German expert committee on nuclear waste site selection process
Andra	French radioactive waste implementer
BECCS	Bioenergy CSS
^{14}C	Carbon isotope 14
CCS	Carbon capture and storage
CCUS	Carbon capture, utilisation and storage
CDM	Clean Development Mechanism
CEC	Commission of European Communities
cf	Compare
CH	Switzerland
CO_2	Carbon dioxide
CO_2e	Carbon dioxide equivalent
CO_2eq	Carbon dioxide equivalent
CoRWM	Committee on Radioactive Waste Management (UK)
DACCS	Direct air CCS
DOE	US Department of Energy
EBS	Engineered barrier system
EC	European Commission
EEA	European Environment Agency
EOR	Enhanced oil recovery
ETH	Swiss Federal Institute of Technology
ETS	EU Emissions Trading System
EU	European Union
EURc	Euro, Euro cent
FADWO	Swiss Federal Ordinance on the Avoidance and the Disposal of Waste
FEPA	Swiss Federal Environmental Protection Act
FOEN	Swiss Federal Office for the Environment
FRPA	Swiss Federal Radiation Protection Act
GDP	Gross domestic product
GHG	Greenhouse gas
GHI	Global Health Index
Gt	Gigatonne (billion metric tonnes)
h	Hour
HLW	High-level radioactive waste
IAEA	International Atomic Energy Agency
ibid./id.	See above, same
IBRD	The International Bank for Reconstruction and Development
ICRP	International Committee on Radiological Protection
IEA	International Energy Agency
ILW	(long-lived) Intermediate-level waste
IPCC	Intergovernmental Panel on Climate Change
IRGC	International Risk Governance Council

kWh	Kilowatt hour
kyr	Kilo year (1000 years)
LCA	Life cycle analysis/assessment
LD-50	Lethal dose LD_{50} (with 50% fatalities)
LLW	Low-level radioactive waste
m	Metre
MCDA	Multi-criteria decision analysis
Nagra	National Cooperative for Radioactive Waste (Swiss implementer)
NEA	Nuclear Energy Agency (of the OECD)
NGO	Non-governmental organisation
NIMBY	Not in my backyard
NIMTOO	Not in my term of office
NL	The Netherlands
NO_x	Nitrous oxides
NPP	Nuclear power plant
NRC	US National Research Council/Nuclear Regulatory Commission
NV	Nevada
NWMO	Nuclear Waste Management Organization (Canada)
OECD	Organisation for Economic Cooperation and Development
R&D	Research and development
RD&D	Research, development and demonstration
SDG	Sustainable development goals
SFOE	Swiss Federal Office of Energy
SFOEN	Swiss Federal Office for the Environment
SKB	Svensk Kärnbränslehantering AB (Swedish Nuclear Fuel and Waste Management Company)
SNEA	Swiss Nuclear Energy Act
S(N)F	Spent (nuclear) fuel
STS	Science and technology studies
SWOT	Strengths, Weaknesses, Opportunities, Threats (evaluation method)
t, tn	(Metric) tonne
TA	Technology assessment
UK	United Kingdom
UN	United Nations
UNDP	United Nations Development Programme
UNEA	United Nations Environment Assembly
URL	Underground research laboratory
US(A)	United States (of America)
USD/US$	US dollar
WHO	World Health Organization
WIPP	Waste Isolation Pilot Plant (nuclear repository in New Mexico, USA)
WRI	World Resources Institute
Y, yr	Year

List of Figures

List of Tables

List of Boxes

Introduction: Setting the Problem(s)

1

Abstract

Deep geological repositories of nuclear waste, long-term landfills of special waste and the substantive storage of carbon dioxide to relieve the world's climate from excessive system change are intricate and contentious policy fields with an impact of decades to hundreds of thousands of years. The present work aims to explore how society and technology can tread and build sustainable paths—in the long run—to cope with the issues.

Keywords

Wicked/messy problems • Environmental problems • Strategic Monitoring • Sociotechnical systems • Long-term safety • Long-term project • Long-term governance • Radioactive/nuclear waste • Conventional toxic/hazardous waste • Carbon Capture, Utilisation, and Storage, CCUS/CCS

Learning Objectives

- Be aware of complex, contentious, long-term problems;
- Become familiar with sociotechnical issues and systems: Recognise the interplay of technical, institutional, social and political aspects;
- Get familiar with the concept of "Strategic Monitoring" of nuclear waste, conventional toxic waste and carbon dioxide storage;
- Realise that you belong to the target audience of this contribution.

1.1 What is the Problem?

(High-level) "nuclear waste management has the deserved reputation as one of the most intractable policy issues facing the United States and other nations using nuclear reactors for electric power generation" (North 1999)—this 20-year old quotation is still valid and no corresponding repository[1] has been built yet or is in operation worldwide.[2] Radioactive waste management has been called a "wicked problem" (Rittel and Webber 1973, Brunnengräber 2019) though "messy" may be a more adequate and less biased term (Metlay and Sarewitz 2012). Such problems are complex, ill-defined or ill-structured, their framing is difficult and there are no clear stopping rules (Reitman 1974); the issue cannot readily be "closed" or (re-)solved (Box 1.1). "Ill-defined" is meant in the sense that there is no single silver bullet or standard solution but there are, e.g., many (national) solutions. Similar situations apply to the management of conventional (non-radioactive) hazardous waste or the "elimination" of excessive CO_2 in the atmosphere (Box 1.2, Fig. 1.1).

> **Box 1.1: What is a Problem, a Complex Problem and an Environmental Problem?**
> *Problems* are defined by the perception of the difference between a final state (sought after) and an actual state (unwanted). Decision problems are well-structured if the decider is familiar with their initial state and the goal state as well as a defined set of transitions. *Complex problems* are intransparent, have multiple goals (called polytely), situational complexity and time-delayed effects (Funke 1991). *Environmental problems* often are complex and often ill-structured or ill-defined (Fig. 1.1).

In addition to the difficulty of problem definition, we must acknowledge that all three policy areas pose so-called "implicit problems", i.e., they were caused by preceding activities or decisions and now constitute (factual) constraints. Academics, consequently, are in the uneasy situation that respective research—in whatever direction—is

---

[1] For key terms and concepts refer to Glossary in the back.

[2] Though some countries are more advanced than others (Sect. 6.2.1).

© The Author(s), under exclusive license to Springer Nature Switzerland AG 2023
T. Flüeler, *Governance of Radioactive Waste, Special Waste and Carbon Storage*, Springer Textbooks in Earth Sciences, Geography and Environment, https://doi.org/10.1007/978-3-031-03902-7_1

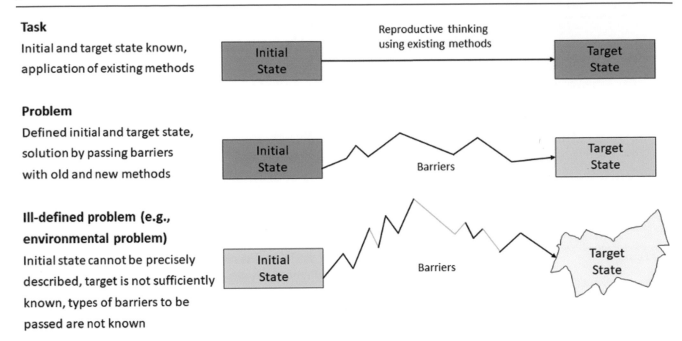

Fig. 1.1 Environmental problems are "ill-defined", with many barriers to a—possibly—known or unknown solution, the target state (after Scholz and Tietje 2002)

"supportive" research. This underlying factual constraint determines the debate and makes it understandable that the issues are politically so contentious.

Box 1.2: Three Case Studies

- Radioactive waste: risk characteristics (Sect. 4.4, Tables 4.2 and 4.5), evolution of risk and safety concepts (Sect. 4.5), system, governance and institutions (Sects. 5.6, 5.7, Tables 4.1 and 4.3);
- Conventional toxic waste[3]: risk characteristics (Sect. 4.4, Table 4.2), evolution of risk and safety concepts (Sect. 4.5), system, governance and institutions (Sect. 5.6, Tables 4.1 and 4.3);
- Carbon Capture ('Utilisation') and Storage, CC(U)S[4]: risk characteristics (Sect. 4.4, Table 4.5), evolution of risk and safety concepts (Sects. 2.3, 4.5), system, governance and institutions (Sects. 2.4, 5.6, Tables 4.1 and 4.4);
- Definitions of waste in Section 4.2, Box 4.1.

---

[3] Nuclear or radioactive waste is also "hazardous" waste. Here "hazardous" or "toxic" or "special" waste is associated with conventional, i.e., non-ionising, long-term waste.

[4] Henceforth, the term "CCS" for Carbon Capture and Storage is used (knowing that utilisation may also be part of it: CCUS) as the final disposition of CO_2 is the focus of this study.

1.2 Bedevilled by the Long-Term Dimension(s)

The hazard potential of toxic conventional and of radioactive wastes is often defined by the toxicity of in part highly concentrated substances. But some ecotoxical danger may also stem from low concentrations. Decisive are the potentially long-term effects of substances, hard if not impossible to forecast and truly steer. Carbon dioxide is detrimental as it accumulates in the atmosphere and, by that, increases the greenhouse effect. In all three cases, the main mechanism of the system is a low-level, but long-term, chronic release into the environment (Chap. 4). Aside from this objective safety aspect, the governance of the three policy areas has a long-term project character: Siting, etc., procedures, knowledge transfer and duties (e.g., on-site monitoring) must be handed over to future generations (Chap. 5). Centuries, even decades, are very long in human terms. Discontinuation must, at any rate, be avoided (Fig. 1.2). This circumstance is the practical legitimation for the present work, a suggestion to use the concept of "Strategic Monitoring"—to oversee the "good governance" of designing and implementing respective programmes of the three waste systems (carbon dioxide also defined as waste in the sense that it is an unwanted substance). All notions will be thoroughly discussed: the "objective" notion of "long term" in Section 4.4 and notions beyond in Section 5.2.

Fig. 1.2 The policy areas under scrutiny are not just "technical" issues but must be dealt with by multiple generations (Fig. 5.1). Passing on the baton is crucial, a failure would be disastrous (Photos by Google image search)

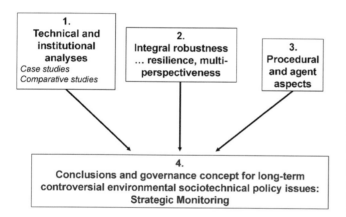

Fig. 1.3 Combined approach to tackle the task "Strategic Monitoring" of the three policy areas under discussion (Annex)

1.3 Approach: Structure of Work, Limitations and Graphical Guide

A complex issue usually needs a complex approach. It is inter-, multi- and even transdisciplinary as it concerns and affects many fields and actors[5] (Box 1.3). The evidence is based both on long-standing research and experience "in practice" in various functions (see About the Author). The presented concept of "Strategic Monitoring", as shown in Fig. 1.3, relies on:

- own extensive technical and institutional analyses in all three policy areas;

---

[5] Parties involved are interchangeably named actors, agents, players, or stakeholders.

- a concept of integral robustness (which is developed below, Sects. 5.1 and 5.3, Flüeler 2006b);
- studies of procedural and agent (actor) aspects.

Box 1.3: Inter-, Multi-, Transdisciplinary … and "Thick" Descriptions

Interdisciplinary studies are executed by two or more autonomous disciplines, disrupting disciplinary boundaries or not, but are mostly object- and/or problem-driven (for a detailed analysis Barry et al. 2008). Climate science per se is inter-, cross- and transdisciplinary—the global environment is not just a system (of systems) but an object of global government or, rather, governance (Schellnhuber 1999). Disciplines have split up (e.g., geography); new research questions created new disciplines (e.g., environmental sciences). *Multidisciplinarity* is characterised by the cooperation of several unchanged disciplines, working within their distinct self-defined framings. Interdisciplinarity may lead to more integrative knowledge or "subordinate" disciplines may serve others (economy and psychology support a novel and/or contested technology, Goldblatt et al. 2012).

Transdisciplinarity transcends disciplinary norms. According to Gibbons and colleagues, problem-driven, dynamic transdisciplinarity is one attribute of the so-called "mode 2" of knowledge production. On top of, or beside, "mode 1" of conventional disciplinary scientific knowledge acquisition, this novel strategy to gain insight into the world is produced in the process-oriented context of application; it is marked by new, often transient, forms of organisation with members of heterogeneous experience: "The experience gathered in this process creates a competence which becomes highly valued and which is transferred to new contexts" (Gibbons et al. 1994, 6; cf. quotes by Fleck in Preface, etc.). By integrating a number of interests, also so-called concerned groups or stakeholders, mode 2 knowledge gains more social accountability and makes all participants more reflexive. Gibbons and colleagues even argue that "the individuals themselves cannot function effectively without reflecting—trying to operate from the standpoint of—all the actors involved" (ibid., 6).

Scholz (2000) qualifies the transdisciplinary approach to expand knowledge generation: "Transdisciplinarity aspires to make the change *from research for society to research with society* … mutual learning sessions … should be regarded as a tool to establish an efficient transfer of knowledge both from science to society and from problem owners (i.e. from science, industry, politics etc.) to science" (Scholz 2000, 13, Italics original). Nowotny and colleagues even talk of "trading zones of knowledge" (Nowotny et al. 2001, 143–147).

Regarding the attempted "sweeping", the all-encompassing approach, it is helpful to call in the method of *"deep descriptions"* from anthropology: "The important thing about the anthropologist's findings is their complex specificness, their circumstantiality. It is with the kind of material produced by long-term, mainly (though not exclusively) qualitative, highly participative, and almost obsessively fine-comb field study in confined contexts that the mega-concepts with which contemporary social science is afflicted—legitimacy, modernization, integration, conflict, charisma, structure, … meaning—can be given the sort of sensible actuality that makes it possible to think not only realistically and concretely *about* them, but, what is more important, creatively and imaginatively *with* them" (Geertz 1973, Italics original).

The documents (with a large body of references) are listed in the Annex.

The reader is guided through the book by the linked graphical flowchart (Fig. 1.4). The complexity of each policy field has to be acknowledged but, at the same time, it must be duly described, analysed and tackled. This is done with the help of various, and diverse, concepts and disciplines:

– All three waste systems are considered *"sociotechnical systems"* (Ropohl 1999) as they are composed of (geo-) technical and institutional (political) subsystems, and that

Fig. 1.4 The concept of "Strategic Monitoring" is built according to the logic below (clickable numbers refer to chapters)

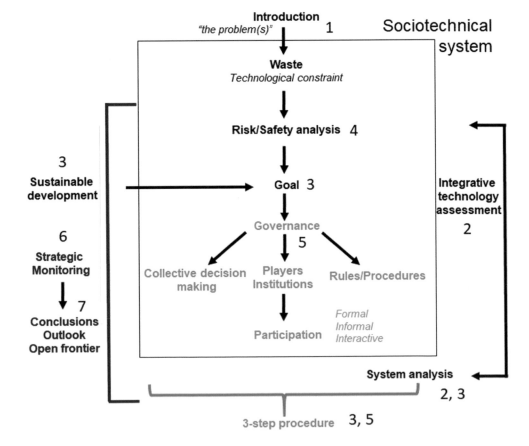

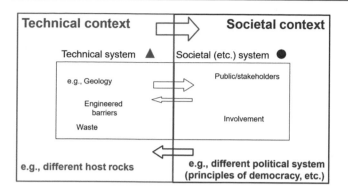

Fig. 1.5 All three waste disposal issues are sociotechnical systems: nuclear or conventional waste or carbon dioxide as technological constraints in the (geo-)technical and sociocultural context. This technological "overprint" is denoted by thicker arrows from left to right (reproduced from Flüeler et al. 2009c)

in a sociocultural context, e.g., with different host rocks and in different political systems and cultures (Fig. 1.5). And they are Large Technical Systems at that (Box 1.4). These circumstances require caution when proposing generic recommendations.

- (Integrative) *technology assessment* gives an appraisal of the overall "super" systems the respective waste systems are part of: the nuclear industry[6] for radioactive waste (NEA 2020), the resource economy for conventional hazardous waste (Baccini and Brunner 2012; EU 2019) and the fossil-based economy for carbon storage (exemplified in Chap. 2 with CCS; IPCC 2005; IEA 2020). This perspective reminds us that each waste system is generated by higher-level systems (and decisions). Waste always is an "implicit" problem with a submissive smack and legacy.
- (Integrative) *risk analysis* describes the technical characteristics, the safety concepts and the institutional setting of each waste system (Chap. 4). The notion of *robustness* is a first attempt to cope with the complexity of the three waste systems.
- Implicit or not, the waste systems do pose problems to be addressed. As they are highly controversial and respective disposal programmes often meet severe opposition, criteria and rules are needed to find some ways out of this situation. The notions of *multiperspectiveness and sustainable development* are proposed to find some "common ground" for solutions or solution spaces (Chap. 3).
- As not only the objective characteristics of the wastes require a long-term perspective (need for isolation, in the order of 10,000 to 1 million years) but also the respective programmes are long-lasting, robustness is complemented by *resilience* to recognise the need for flexibility and

discourse over decades. Mature visions of *governance and institutional settings* are required (Chap. 5).

- Addressing the institutional long term (in the order of decades), "Strategic Monitoring" is suggested to see that the corresponding programmes are on track and duly implemented (Chap. 6).
- Upon the bold idea of "Strategic Monitoring", more modest (truly?) conclusions follow, i.e., to become long-term literate, as persons, experts, laypeople, as well as societies, to truly adopt some ownership of the problem—"waste" (Chap. 7).

Box 1.4: Large (Socio)Technical Systems
A *sociotechnical system* is "an action or working system where human and technical subsystems constitute a unity" (Ropohl 2009, 141, transl. tf).

Large technical systems are mainly infrastructure systems with specific functions (e.g., electricity distribution), particular knowledge base and norms as well as specialised professional groups. They are characterised by a high degree of technical and social integration. Their interconnectedness may yield to more technical disruption and/or political interference. They should not be confounded with large-scale technical projects like the Manhattan project to construct an atomic bomb. Waste management and disposal are typical infrastructure systems, with complex technical components and organisational networks subject to a high degree of regulation (by state laws and/or technical norms). See, e.g., Hughes (1983), Joerges (1988) and Mayntz (1988).

Finally, it is important to note that the starting point of this undertaking and the author's core competence is the policy field of radioactive waste, knowing the international scene but digging from the Swiss case. With due reference to other national programmes, these insights and proposals were extended to and reflected upon two other policy fields, conventional hazardous waste and carbon storage. Experts in these areas will hopefully not hold shortcomings on the author's side against him but, on the contrary, take this as food for thought, reflection and development—not as a model—in the fields; they know better than the author does.

1.4 Audience: Both Practitioners and Advanced Students

The approach provides all necessary background information to be sensibly used. Readers need not be specialists in a particular field; therefore, the present guide is appropriate for

---

[6] The weapons and, generally, military origin of the issue is left aside in the following considerations.

various audiences—practitioners, policy makers, research policy makers, researchers, as well as (technical) graduates and (non-technical) undergraduates (technical undergraduates who supposedly have no time to already cover such issues in their full complexity). Respective knowledge and skills for "handing over the torch" are so far neither part of academic curricula nor of competences in industrial companies or government services.

This book is a risk. It attempts to present three "messy" complex and contentious issues in their technical, social and political context, with the aim to propose some guidance for a respective "long-term governance". The format is a text-book, but an exploratory one at that. It is not a clear-cut instruction manual, often applied with craftsmanship and with success. It is a tentative proposal to learn from "messy" experience (Annex) and responsibly find our ways through the swamp called "sustainable future".

The marketing strategy to make the book accessible on a chapter-by-chapter basis implies some redundancies. As the disciplines drawn on are manifold, the references are like-wise. They are classified in "key references" of the author's own and basic publications categorised in academic fields or topics underneath. "Sources" indicate where quotes, defini-tions, information in boxes come from. First readers may limit themselves to key references; further literature struc-tured by theme enables a focused in-depth study. More underlying references are contained in the author's reference list in the Annex. In order not to present a reference overkill, confirmatory more recent literature is only mentioned selectively.

Future

> L'avenir n'est jamais que du présent à mettre en ordre.
> Tu n'as pas à le prévoir mais à le permettre.
> *The future is never more than the present to put in order.*
> *Your task is not to foresee it, but to enable it.*

Antoine de Saint-Exupéry (1900–1944), Citadelle, 1948

1.5 Conclusion and Summary

Deep geological repositories of nuclear waste, long-term landfills of special waste and the substantive storage of carbon dioxide to relieve the world's climate from excessive system change are intricate and contentious policy fields with an impact of decades to hundreds of thousands of years. This work aims to explore how society and technology can set up and implement sustainable ways—in the long run—to cope with the issues. They all are long-term safety issues and require long-term, decades-long institutional involvement of the technoscientific community, waste producers, public

administrators, non-governmental organisations, NGOs, and the public. The demonstration of long-term safety is chal-lenging, and monitoring may contribute to substantiate evi-dence, support decision making and legitimise the programmes. What, where and when to monitor is deter-mined by its goal setting: it may be operational, confirmatory (in the near field) or environmental (far field) (Fig. 6.3). As it is "difficult to make predictions, especially about the future",[7] this contribution, too, does not pretend to present the silver bullet, but Strategic Monitoring as proposed con-tributes to process, implementation or policy and institu-tional surveillance to sustain a once launched programme. It addresses not only the controversial long-lasting "problem" (of nuclear, toxic or CO_2 waste) but also investigates some ways to approach "solutions" or solution spaces, not only technical but also societal, institutional and personal, and this is for the long term. It includes the tailored transfer of knowledge, concept and system understanding, experience and documentation to various audiences above. It is an integrative tool of targeted, yet adaptive, management and may be applicable to other long-term sociotechnical fields.

Questions

1. Why are radioactive waste, conventional toxic waste and Carbon Capture and Storage, CCS, "implicit" problems?
2. Why is it "rational" to link the issue of radioactive waste with the operation of nuclear power reactors?
3. Why is it also "rational" to decouple the issue of radioactive waste from the generation of electricity by nuclear power plants?

Answers

1. Radioactive waste is generated by the utilisation of radionuclides in power plants, medicine, industry and research. Conventional waste comes from the use and abandonment of materials and goods in economy (from industry to households). CCS is propagated as a possible "solution" for the massive burning of fossil fuels (for heating, energy production or automobiles).
2. Most radioactive waste stems from the fuel burnt in nuclear power reactors and their decommissioning.
3. Radioactive waste as a "technological constraint" (from nuclear power plants, etc.) exists … and has to be handled in a sustainable way, irrespective of its origin.

---

[7] Famous saying of Danish origin, proverb or Niels Bohr, also attributed to Mark Twain, Samuel Goldwyn or Nostradamus (https://quoteinvestigator.com/2013/10/20/no-predict/, accessed 27 January 2023).

Additional Information

All weblinks accessed 27 January 2023.

Key Readings

Problems

Brunnengräber A (2019) The wicked problem of long term radioactive waste governance. In: Brunnengräber A, Di Nucci MR (eds) Conflicts, participation and acceptability in nuclear waste governance. An international comparison, vol 3. Springer, Wiesbaden, pp 335–355 (wicked problems)

Funke J (1991) Solving complex problems: exploration and control of complex systems. In: Sternberg R, Frensch F (eds) Complex problem solving—principles and mechanisms. Lawrence Erlbaum Associates, Hillsdale, NJ, pp 185–222 (complex problems)

Metlay D, Sarewitz D (2012) Decision strategies for addressing complex, "messy" problems. The Bridge—linking engineering and society. Social Sciences and Engineering Practice. Fall 2012, 42(3):6–16 (The National Academy of Engineering. National Academy Press, Washington, DC). https://www.nae.edu/62558/Decision-Strategies-for-Addressing-Complex-Messy-Problems (messy problems)

Reitman WR (1964) Heuristic decision procedures, open constraints, and the structure of ill-defined problems. In: Shelly II WM, Bryan GL (eds) Human judgments and optimality. Wiley, New York, pp 282–315 (ill-defined problems)

Rittel HWJ, Webber MM (1973) Dilemmas in a general theory of planning. Policy Sci 4:155–169. https://doi.org/10.1007/BF01405730 (wicked problems)

Scholz RW, Tietje O (2002) Embedded case study methods. Integrating quantitative and qualitative knowledge. Sage Publications, Thousand Oaks. https://doi.org/10.4135/9781412984027 (environmental problems)

Systems

Flüeler T (2006b) Decision making for complex socio-technical systems. Robustness from lessons learned in long-term radioactive waste governance. Environment & Policy, vol 42. Springer, Dordrecht NL. https://doi.org/10.1007/1-4020-3529-2

Hughes TP (1983) Networks of power. Electrification in western society 1880–1930. Johns Hopkins University Press, Baltimore MD

Joerges B (1988) Large technical systems: concepts and issues. In: Mayntz R, Hughes TP (eds) The development of large technical systems. Campus, Frankfurt am Main/Westview, Boulder CO

Mayntz R (1988) Zur Entwicklung technischer Infrastruktursysteme. In: Mayntz R, Rosewitz B, Schimank U, Stichweh R (eds) Differenzierung und Verselbständigung. Zur Entwicklung gesellschaftlicher Teilsysteme. Campus, Frankfurt am Main, pp 233–259

Ropohl G (1999) Philosophy of socio-technical systems. Soc Philos Technol Q Electron J 4(3):186–194. https://doi.org/10.5840/techne19994311

Ropohl G (2009) Allgemeine Technologie. Eine Systemtheorie der Technik. Universitätsverlag Karlsruhe, Karlsruhe

From Inter- to Transdisciplinarity

Barry A, Born G, Weszkalnys G (2008) Logics of interdisciplinarity. Econ Soc 37(1):20–49. https://doi.org/10.1080/03085140701760841

Gibbons M, Limoges C, Nowotny H, Schwartzman S, Scott P, Trow M (1994) The new production of knowledge. The dynamics of science and research in contemporary societies. Sage, London. https://doi.org/10.4135/9781446221853

Goldblatt DL, Minsch J, Flüeler T, Spreng D (2012) Introduction (Chap. 1). In: Spreng D, Flüeler T, Goldblatt DL, Minsch J (eds) Tackling long-term global energy problems: the contribution of social science. Environment & Policy, vol 52. Springer, Dordrecht NL, pp 3–10. https://doi.org/10.1007/978-94-007-2333-7_1

Nowotny H, Scott P, Gibbons M (2001) Re-thinking science. Knowledge and the people in an age of uncertainty. Polity Press, Cambridge UK. https://www.wiley.com/en-us/Re+Thinking+Science%3A+Knowledge+and+the+Public+in+an+Age+of+Uncertainty-p-9780745626079

Schellnhuber HJ (1999) 'Earth system' analysis and the second Copernican revolution. Nature supplement, 402(2 Dec):19–23

Scholz RW (2000) Mutual learning as a basic principle of transdisciplinarity. In: Scholz RW, Häberli R, Bill A, Welti M (eds) Transdisciplinarity: joint problem-solving among science, technology and society. In: Proceedings of the International Transdisciplinarity 2000 Conference, Zurich, 27 Feb–1 March. Workbook II: Mutual learning sessions, vol 2. Haffmanns Sachbuch, Zürich, pp 13–17

Nuclear Waste

NEA, Nuclear Energy Agency (2020) Management and disposal of high-level radioactive waste: global progress and solutions. OECD/NEA, Paris. https://www.oecd-nea.org/jcms/pl_32567

Conventional Toxic/Hazardous Waste

Baccini P, Brunner PH (2012) Metabolism of the anthroposphere. Analysis, evaluation, design. MIT Press, Cambridge MA. https://mitpress.mit.edu/9780262016650/metabolism-of-the-anthroposphere/

EU, European Union (2019) https://ec.europa.eu/environment/waste/target_review.htm

Carbon Capture ('Utilisation') and Storage, CC(U)S

IEA, International Energy Agency (2020) Energy technology perspectives 2020. Special report on carbon capture, utilisation and storage. CCUS in clean energy transitions. IEA, Paris, 174 pp. https://iea.blob.core.windows.net/assets/181b48b4-323f-454d-96fb-0bb1889d96a9/CCUS_in_clean_energy_transitions.pdf

IPCC, Intergovernmental Panel on Climate Change (2005) IPCC special report on carbon dioxide capture and storage. In: Metz B, Davidson O, de Coninck HC, Loos M, Meyer LA (eds) Prepared by working group III. Cambridge University Press, Cambridge UK. https://www.ipcc.ch/report/carbon-dioxide-capture-and-storage/

Websites

https://www.iaea.org/topics/radioactive-waste-and-spent-fuel-management (International Atomic Energy Agency: Radioactive waste)

https://www.oecd-nea.org/jcms/c_12892/radioactive-waste-management/ (Nuclear Energy Agency of the OECD: Radioactive waste)

https://www.eea.europa.eu/themes/waste (European Environment Agency: Conventional waste within European Union)

http://www.basel.int/TheConvention/Overview/tabid/1271/Default.aspx (Basel Convention on the Control of Transboundary Movements of Hazardous Wastes and their Disposal)

https://www.ipcc.ch/ (Intergovernmental Panel on Climate Change: Carbon storage)

Sources

Flüeler T, Stauffacher M, Krütli P, Moser C, Scholz RW (2009c) Cross-cultural space of radioactive waste governance in a globalizing world. Presentation at Panel 2 "The role of geography". Conference "Managing Radioactive Waste Problems and Challenges in a Globalizing World", Gothenburg, 15–17 Dec 2009 (slide #11) (Fig. 1.5)

Geertz G (1973) Thick description. Toward an interpretive theory of culture. In: The interpretation of cultures: selected essays. Basic Books, New York, pp 3–30. http://hypergeertz.jku.at/geertztexts/thick_description.htm, from: http://xroads.virginia.edu/~DRBR/geertz2.txt (Box 1.3)

North DW (1999) A perspective on nuclear waste. Risk Anal 19(4): 751–758, 751 (quote, Sect. 1.1)

Saint-Exupéry A (1948) Citadelle [The wisdom of the sands]. Gallimard, Paris (quote, Sect. 1.4)

Scholz RW, Tietje O (2002) Embedded case study methods. Integrating quantitative and qualitative knowledge. Sage Publications, Thousand Oaks, USA, p 26 (Fig. 1.1)

Integrative Technology Assessment— Proposal for a Framework

2

Abstract

Before needs and goals can be addressed (Chap. 3), an inventory of the total technology system in question—call it stocktaking—must be made. Carbon capture and storage, CCS, highlights the tension between the advantage of a short-term "quick fix" and the disadvantage posed by the risk of long-term leakage and, from a technology policy perspective, the danger of perpetuating carbon lock-in. In an exemplary way, this chapter assesses CCS against criteria taken from the controversial and long-lasting governance of radioactive waste. As the dimensions covered by this issue are manifold and intertwined, there is no "one" methodology with which to analyse it (such as a technology assessment of the n-th order). Instead, cross-disciplinary investigations make it possible to draw lessons from contentious long-term environmental issues and social science research, which is necessary before embarking on this route on a large scale. The outcome is a proposed framework for a system analysis (Chap. 5), as a base for the envisaged Strategic Monitoring (Chap. 6).

Keywords

Technology assessment • Strategic Monitoring • Socio-technical systems • Technical fix • Lock-in • Radio-active/nuclear waste • Conventional toxic/hazardous waste • Carbon Capture (Utilisation) and Storage, CCUS/CCS

Learning Objectives

- Know what technology assessment, TA, is all about;
- Recognise TA as the basis for a comprehensive system analysis and a cornerstone of the envisaged Strategic Monitoring;
- Identify the multiple levels to deal with in the contentious policy field of CCS and beyond.

2.1 How to Assess a Complex Technology?

Although both science and society have recognised climate change as a serious problem[1] (EC 2009; EU 2019; Pew Center 2019), mankind has continued along with a high greenhouse gas, especially carbon dioxide (CO_2), emissions path to the atmosphere. Barriers to effective reductions exist at political, institutional and individual levels. Different parts of the world have diverging views on the responsibilities of "first movers" (historical polluters such as North America, Australia or Europe versus "new" polluters like China, India and developing countries). Incentives, trading and enforcement mechanisms are weak or not in place in these latter countries, and large-scale lifestyle changes towards sustainable development are rare.

In such a tangled situation, the characteristics of CCS seem attractive. It is a quick and technical, narrowly located but a high-potential solution that has no need for widespread and extensive efficiency improvement in dispersed facilities, equipment, appliances or "software" such as institutions and behaviour. In view of the widespread implications, a mere "technical fix" should be avoided, a choice often made in the past (e.g., with pesticides in food production, or nuclear, Metlay 1978) and a label misused in the present, e.g., "climate fix".[2]

> Wen ale mentschn soln zien oif ejn sajt, wolt sich di welt ibergekert.
> *If all men headed for the same direction the world would overturn.*
>
> Yidishe shprikhverter un redensartn. Frankfurt aM, 1908

Whether or not it is in fact "one of the most intractable policy issues" (North 1999, 751) in countries with nuclear

---

[1] For key terms and concepts refer to Glossary in the back.

[2] See Pielke (2010): "The climate fix: what scientists and politicians won't tell you about global warming", an attempt to discredit climate science. For an analysis of such endeavours, read Oreskes and Conway (2010).

T. Flüeler, *Governance of Radioactive Waste, Special Waste and Carbon Storage*, Springer Textbooks in Earth Sciences, Geography and Environment, https://doi.org/10.1007/978-3-031-03902-7_2

power plants, the long-term governance of high-level radioactive waste remains unsolved in most regions of the world and is a major stumbling block for the nuclear industry. Based on the 70-year history and continual evaluations of how industry, governments and society deal with nuclear waste (e.g., Colglazier 1982; Kasperson 1983; Carter 1987; Blowers et al. 1991; COWAM 2003; Freudenburg 2004; Flüeler 2006b; Solomon et al. 2010, many later on, e.g., ARGONA 2017, Sects. 5.6 and 6.2), it is possible to draw some key lessons for other long-term sociotechnical environmental issues such as CCS (also toxic waste in a second round). The approach is to assess the issue using a combination of disciplines and perspectives. These can include systems theory, integrated risk assessment, social sciences, technology assessment (Box 2.1) and management (implementation, compliance) (Flüeler 2006b). The six criteria below address issues that have proven to be crucial in technology policy debates (Ravetz 1980; Vlek and Stallén 1981; Wynne 1980; Morone and Woodhouse 1986; Kasperson et al. 1992; Ropohl 1999, approach in Flüeler 2006b):

1. Need for deployment and comparative benefits;
2. Total-system analysis and safety concept;
3. Dedicated internationally harmonised regulation and control;
4. Economic aspects (costs and incentives);
5. Implementation;
6. Societal issues.

> **Box 2.1: Technology Assessment**
>
> Technology assessment is a scientific, interactive and communicative process which aims to contribute to the formation of public and political opinion on societal aspects of science and technology (Europ. Akademie 2004).
>
> TA as an act of balance in the conflict areas of widening and closing down of options, of gaining autonomy and adjustment needs, of desired and unwanted use, of perspectives of deciders vs. affected parties and of technocracy vs. democracy (Grunwald 2010, 5).

It is important to note that rather than simply being items that are checked off on a list (first "technical", then social "add-ons"), the criteria should relate to one another such that numbers 1 and 6 above are interlinked and influence the others all along. CCS is not "just" a technical innovation but a sociotechnical system that is eventually to be implemented on a large scale. As the CCS concept is a technological system "in the making" (Latour 1987), approaches in science and technology studies, STS, such as "interpretative

flexibility" (with different understandings of the same technology by different people), are useful for scrutinising arguments for their consistency (Collins 1981; Bijker et al. 1987). The following is a suggestion to carry out such an attempt at integral appraisal as well as to create a sound basis for criteria for a Strategic Monitoring (Chap. 6).

2.2 Criterion #1: Need for Deployment and Comparative Benefits Vis-à-Vis Competing Technological Options

Radioactive waste is a technological and political constraint. It results from politically desired and officially promoted activity (the choice of nuclear power in national energy policy as well as from medicine, industry and research), and it is a physical by-product that must be dealt with in a sustainable way. Waste is the inevitable "back end" of the nuclear fuel "cycle" and literally "downstream" in the controversial nuclear debate. It is an intrinsic characteristic of the nuclear system and must be dealt with accordingly.

With the climate debate where it was recognised that CO_2 is a pollutant, this otherwise harmless gas, needed for plant growth, has become waste and part of the back end of fossil fuel systems. As such, it is as inseparable from, i.a., coal utilisation as radioactive waste is from nuclear fission utilisation. There has also been a mention of **closing the carbon cycle** (Powicki 2007) just as the intention to close the nuclear cycle (NEA 2000). Unlike nuclear waste, however, CO_2 is not tied to a specific technology but is produced in conversion chains other than fossil systems (like deforestation) and can be reduced in such chains (such as afforestation with entrained CO_2 sequestration) or in processes (such as building insulation with associated avoided CO_2—negative—emissions). Consequently, CCS is in competition with other technologies with regard to other criteria, such as potential, efficiency and cost, and it must, therefore, demonstrate its necessity and perform uncontrovertibly well. At this stage of development (Research and Development, R&D; few pilots; concept not established; see below), the performance of CCS has yet to be determined.

2.3 Criterion #2: Total-System Analysis and Safety Concept

Even though the approach sounds thrilling as a technical solution, the Carbon Capture and Storage concept is not yet mature (Chaps. 4 and 5), and its technical potential cannot be reliably defined at this stage. Furthermore, its performance cannot be assessed because overall conditions, including its embeddedness in economic, technological and environmental policies, must be set beforehand. It is essential to start at the beginning with a total-system analysis.

System Approach

CCS is not only a potentially powerful but also an ambiguous lever in the transition away from carbon-intensive economies; therefore, if it is ever deployed, it must be done so very prudently. This must start with a careful system analysis in order to understand all possible structural system changes its full-scale implementation might induce. For all energetic technological lines of CCS (pre-combustion, post-combustion and oxyfuel), a comprehensive life cycle analysis along the process chain of extraction of the energy carrier, transport, conversion, CO_2 capture, compression, transportation and storage phases shows a 10–40% increase in resource use (e.g., coal and mining) if a "low-carbon" use of fossil energy is targeted (IPCC 2005; Wuppertal Institute et al. 2008; Grünwald 2008). This implies a decrease in system efficiency to approximately 65%, currently even 45%, and an increase in classically toxic environmental impact (NO_x, heavy metals and fine particles; see Koornneef et al. 2008).

The net greenhouse gas reduction amounts to 78% for a German brown-coal plant (Wuppertal Institute et al. 2008). Constraints on transportation distance and means are required (such as power plants as close as possible to storage sites, pipelines or large tankers). In a long-term management plan, all components have to be set in place in a timely manner so that coal provisioning, plant lifetime and storage volume and facility are synchronised. The German case shows: If power plants were to reach foreseen termination before 2020 (when full-scale CCS might be in operation) and must be replaced without interruption, a smooth transition to low carbon with capture-ready plants would not be and, in actual fact, was not possible (ibid.). "Technical lessons learned" are not enough (EU CO_2 Network 2004)—after all, it is the maturity and deployment of the overall system, not just its components, that counts. The definition of system boundaries and the choice of decisive energy penalty values are crucial, especially given the slim publicly available detailed database (Page et al. 2009, still true in 2022: https://co2re.co/, Sect. 5.6), as the plant efficiency sinks with carbon capture.

It is widely recognised that **CCS** can only have a **bridging function** (de Coninck 2008). In other words, on the way to sustainable energy systems, geoengineering tools (Box 2.2) must not substitute or hamper the development of renewables and efficiency measures because fossil fuels are finite, as is geological storage, despite the possible existence of huge reservoirs. The crucial question is the length of the transition period that CCS could or should be given (or, rather, that the world energy system should be given with the CCS tool) until the combination of efficiency measures and renewables takes effect. Arguably, it takes less time and effort to "add" CCS to the existing energy supply infrastructure than it does to change the overall energy system (provision, distribution and consumption) to be based on renewables and energy efficiency. Some components, drivers and interrelations are noted in Fig. 2.1.

All in all, the system analysis must consider material and energy fluxes (as well as related stocks), the link of CCS with other energy (sub)systems, integration into the market and the interplay of actors (this is a very intricate issue, as alluded to in Praetorius and Schumacher 2009, for example). CCS is "more than a strategy for 'clean coal'" (IEA 2009c, 4); its technology must also be adopted by biomass and gas power plants, and "mixed" portfolios (namely coal/biomass/hydrogen) are possible. Consequently, "negative emission scenarios" that provide a net reduction of CO_2 must be taken into account. The carbon abatement cost curves (Figs. 2.2, 2.3 and 2.4) put it in a nutshell: Conservation (efficiency) measures offer more for low-cost decarbonisation with less uncertainties but their curves steepen exponentially beyond the mid-point; sequestration (especially direct air carbon capture), however, has great uncertainty but provides an enormous long-term potential.

Der ssejchl is a kriecher.
Reason creeps.

Yidishe shprikhverter un redensartn. Frankfurt aM, 1908

In order to increase transparency and comparability, it might be worthwhile to include CCS in a global warming and environmental production **efficiency ranking** (a metric proposed by Feroz et al. 2009). The system approach reveals how its compartments, functions and dynamics are related, whether this is optimised and, if so, to what goal. One outcome of a system analysis is the revealing of **unintended consequences.** An example in the radioactive waste field is the dysfunctions (leakage) of Asse, a German nuclear low-level waste repository (Sect. 5.7; Ilg et al. 2017). To a large extent, this leakage[3] can be traced back to the fact that this site was not designed for waste disposal but was misused as such after it had been exploited for rock salt. This process has had major repercussions on the German disposal programme, incidentally built on the Gorleben site with the same host rock as found in Asse. Consequently, abandoned oil and gas fields have to be carefully evaluated as potential CCS deposit sites, particularly as their deployment is very attractive to exploiters in the form of profits from extraction,

---

[3] Actually, it is an intrusion of water from the neighbouring rock into the Asse salt mine.

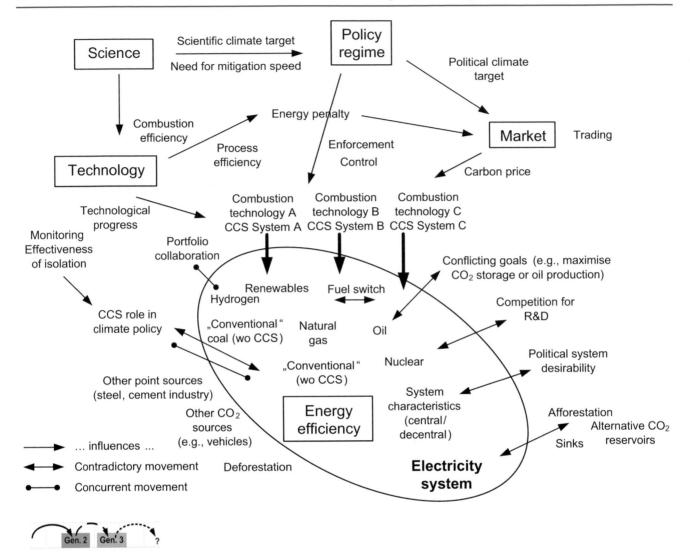

Fig. 2.1 System components, drivers, actors and their interrelationships in the CCS context. Various CCS systems (A, B, C …) depend on science, technology, policy decisions and market conditions (thin arrows). They compete with each other (bold arrows) and interact with other energy and production systems (such as renewables and cement industry, double arrows) (reproduced from Flüeler 2012d)

Fig. 2.2 Drivers for decarbonisation are technological innovation and the economies of scale for both conservation and sequestration paths. Sequestration has a high greenhouse gas (GHG) abatement potential, albeit with a high cost uncertainty, and only if it is successful, i.e., CO_2 stays isolated from the atmosphere (reproduced from Goldman Sachs 2019)

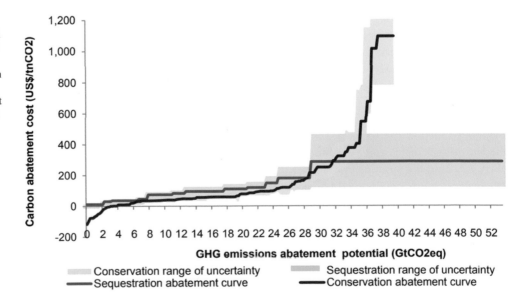

higher yield due to CO_2 injection (up to three times), CO_2 credits and disclaimers in the case of short-term liabilities.

Systemic Features

Some observers have argued that CCS is by no means a novel technological concept as CO_2, among other things, has long been injected into oil formations to recover additional oil (as Enhanced Oil Recovery, EOR, Benson 2005). What is radically different are the respective goals and scales of CCS and EOR. The new intention with CCS is to seclude CO_2 for the common good (a sustainable climate, as it is harmless to individuals, even at high concentrations), at an ample scale and for a very long time. This requires the implementation of a globally effective system, testing long-term safety performance of facilities and the overall system (including an adequate control mechanism) and, above all, the recognition of longevity as a factor to be dealt with.

When CO_2 is emitted, much of it remains in the atmosphere for around 100 years. The carbon cycle of the biosphere takes a long time to neutralise and sequester anthropogenic CO_2. According to Archer, the lifetime of fossil fuel CO_2 is 30,000–35,000 years (Archer 2005) if averaged out over its entire distribution. The term **"lifetime"** refers to the period required to restore equilibrium following an increase in the gas's concentration in the atmosphere. If one includes the long tail, Archer suggests that this figure is "300 years, plus 25% that lasts forever" (ibid., 6). This, along with the heat capacity of the oceans, means that climate warming is expected to continue beyond 2100, even if CO_2 emissions cease.

Comparing radioactive wastes with CO_2 reveals both differences and similarities, both in systemic and risk aspects (Chap. 4, Table 4.4 and Table 4.5, respectively).

Safety Assessment

Safety analyses, which are usually called "performance assessments" in the context of radioactive waste, include scenario development, conceptual and mathematical models development, consequence analysis, uncertainty and sensitivity analysis and confidence building (Chap. 4).

Site Selection and Characterisation Procedure

By definition, disposal facilities also concentrate hazards, which means that the selection of suitable sites is decisive. In the case of radioactive waste, in 2008 the Swiss government started a stepwise site-selection procedure in order to find a suitable site, for all waste types, within 10 years. Although this may not be the case in every country, efforts have been made in a range of national programmes to establish a systematic, transparent and participatory concept for site selection (these include AkEnd for Germany, NWMO for Canada and CoRWM for the UK, Chap. 5 and Sect. 6.2).

Box 2.2: Geoengineering

Methods aim to deliberately alter the climate system to counter climate change, esp. solar radiation management and carbon dioxide removal (IPCC 2013, WG1AR5. Summary for Policy Makers, 29).

Ban Ki-moon on Geoengineering

New technologies to combat global warming could complement reductions in greenhouse-gas emissions. But their potential impact is highly uncertain, and failure to govern their use properly could aggravate existing threats to international peace and security (...).

But at present, we have no idea what unforeseen and unintended consequences deploying these new technologies might have. The unknown unknowns – especially with solar geoengineering – could be just as bad as the known challenges presented by climate change.

What's more? As with global warming, the impact of these technologies will transcend national borders. This puts those who have the least say – the vulnerable and the poor – on the front line. It also risks exacerbating wider threats to international peace and security, such as resource scarcity and forced climate migration.

This is why the Elders, a group of independent global leaders, is calling on the international community to agree on a rigorous governance framework for geoengineering and to put it in place without delay. Such a decision-making system must be transparent, participatory and accountable. It should include the voices of those most affected and enable all governments and non-governmental stakeholders to gain the fullest possible understanding of these new technologies for more informed decision making.

Since the Industrial Revolution, we have known that technology is not a panacea and that it advances human wellbeing only if all those affected are given the chance to participate in its development. This consideration is all especially relevant to geoengineering, because our knowledge of these technologies and their impact remains limited.

Fortunately, efforts are underway to address this. This week, the UN Environment Assembly – the world's highest-level decision-making body on environmental issues – will consider whether to initiate a global learning process on both the science and governance of geoengineering. To this end, the UNEA would call for a worldwide assessment of these emerging technologies, giving all countries a common platform of knowledge.

This shared understanding is an important first step toward ensuring that decisions concerning the use or non-use of geoengineering are based on the principles of equality, justice and universal rights. These are the values underlying the 2015 Paris climate agreement and the Sustainable Development Goals, both of which were adopted during my tenure as UN Secretary-General.

The UN is best placed to accommodate the governance framework requirements we now need. Only through the UN's multilateral processes can we ensure that geoengineering technologies, and how they might be applied, are not the preserve of individual states. This is vital for environmental sustainability, international security and the wellbeing of future generations around the world.

Many people are wary of this debate, particularly in international fora. They fear that it could be a foot in the door for highly dangerous ideas and that the very act of drawing attention to these technologies could reduce pressure to cut emissions.

I understand these concerns, and I agree that our main collective priority must still be to cut emissions; end the use of fossil fuels; and promote a zero-carbon, climate-resilient and people-centred economic transition.

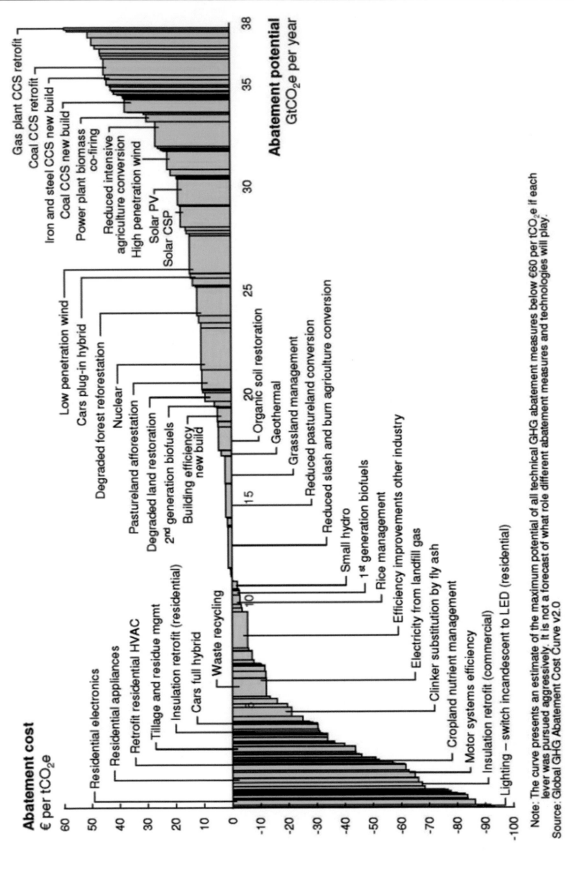

Fig. 2.3 Global greenhouse gas (GHG) abatement cost curve for 2030 and beyond business as usual. The investment costs for any CCS system are high as is their share in CO_2 abatement potential (reproduced from McKinsey 2009 in Flüeler 2012d)

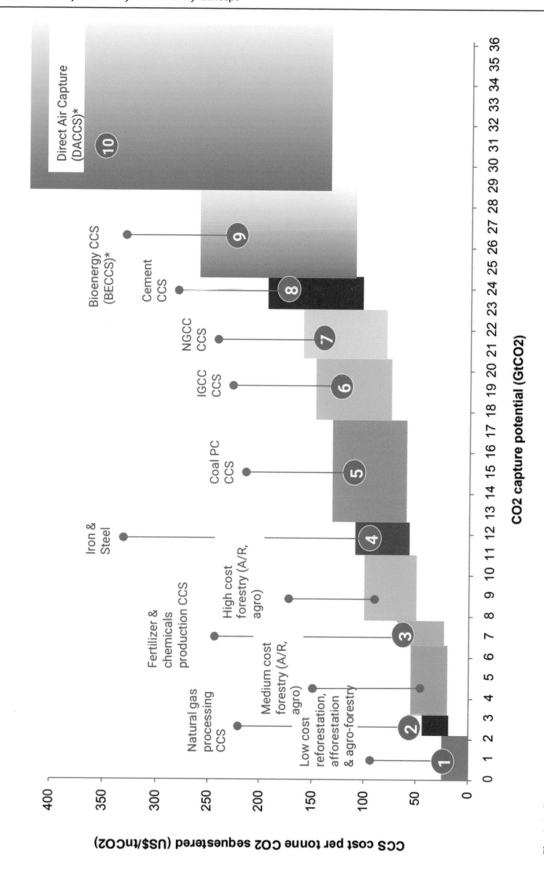

Fig. 2.4 Carbon sequestration includes several carbon capture techniques, bioenergy CCS and direct air capture, DACCS having both the greatest potential and the most uncertainty. Note the currency differing from Fig. 2.3 (reproduced from Goldman Sachs 2019). IGCC Integrated Gasification Combined Cycle; NGCC Natural Gas Combined Cycle

But we also need to acknowledge that the geoengineering genie is already out of the bottle. The likelihood of unilateral deployment of solar geoengineering increases every year. The global community must decide whether to engage now, by setting clear governance rules and guardrails, or allow individual actors to take the lead, creating a fait accompli for the rest of us.

Ignoring this debate would be a mistake. Instead, the world should focus on learning more, including via the process at the UNEA, in order to understand the full range of options and assess their risks with the best information available.

How to understand and potentially harness disruptive new technologies for the benefit of all humanity is one of the defining questions of our age. Future generations will not forgive us if we fail to answer it convincingly.

Ban Ki-moon, Deputy Chair of The Elders, Secretary-General of the United Nations from 2007 to 2016, statement in view of the UN Environment Assembly in March 2019 (Ki-moon 2019).

No progress has been made since (Geneva Environment Network 2022).

2.4 Criterion #3: Dedicated Internationally Harmonised Regulation and Control

Because CCS is seen as part of the mitigation portfolio addressing climate change, straightforward international regulation is pivotal. In fact, it is a prerequisite before its effective deployment can take place (Mace et al. 2007; Steeneveldt et al. 2006). Basic research (for example, on modelling, system definitions, side effects and cost-benefit analyses of the heterogeneous basket of ideas in geoengineering) requires massive but standardised resources. Applicants of non-commercial activities (such as demonstration projects) need funds, while affected stakeholders and non-governmental organisations need confidence in effective regulators. Monitoring and verification depend on international standards, transboundary transfer and trading require the existence of international inventories and transnational treaties and liability issues need to be regulated. The 2009 IEA Roadmap proposed such comprehensive frameworks by 2020 (IEA 2009c, 36–37) and demands (Gale and Read 2005), while recommendations exist in rather generic forms (IRGC 2007, 2008; WRI 2008). Such steps revert to the criteria above; in order to integrate CCS into the Clean Development Mechanism, CDM, for example, technical and organisational aspects must be fully covered, in the form of thorough site-specific risk assessments, good site characterisations and appropriate monitoring during and after operation (Dixon 2009), including corresponding accounting (Haefeli et al. 2005). Decisions are critical in cases where host country regulatory capacities are insufficient. In such cases, Pollak and Wilson suggested giving "preference to sites that present lower seepage risk, such as offshore sites or sites that have well-characterized geology" (Pollak and Wilson 2009, 84). They rightfully concluded that "if a geological storage site approved by the

CDM were to cause environmental, health or safety problems for the host country, it would damage not just the reputation of the CDM but also the reputation of CCS as a greenhouse gas mitigation technology" (ibid., 85). The issues appear complicated in terms of attempts to set up multinational repositories for radioactive waste (IAEA 1998, 2004). For the speedy set-up of a comprehensive, yet effective, functional framework, early abatement is judged less costly and more effective than postponement (Stern 2007; Krey and Riahi 2009). This appraisal holds true at this stage (IEA 2021).

If CCS is to be a plausible part of the mitigation portfolio addressing climate change, "control" will also have to include measures for enforcement. CCS is likely to be so costly that it could be tempting to agree to a range of international treaties and then delay their implementation. Before embarking on the road to full-scale CCS, **rules and institutions** must be put in place that provide reasonable assurance that treaties can be enforced and project targets met. The financial crisis of 2007–2008 in the European Union clearly shows that a functioning mechanism for enforcing regulation must be part of any plausible regime.

2.5 Criterion #4: Economic Aspects (Costs and Incentives)

Costs depend on technology and a range of externally set assumptions (including size, inception, promotion strategies in early phases and coverage of carbon penalty). Differences in abatement technology lines range from −100 EUR/t CO_2 avoided (for high-tech lighting) to +60 EUR/t CO_2 avoided in the case of a retrofitted gas power plant with CCS, as compiled by McKinsey (2009) (Fig. 2.3) and updated by Goldman Sachs (2019) (Fig. 2.4).

The abatement potential for CCS is substantial, albeit at a high cost, according to energy carrier, from 30 (coal) to 55–60 EUR/t CO_2 avoided (gas, retrofitted). Goldman Sachs gives a range even for coal of around 60–130 USD/t. One asset of CCS is the fact that it can also be used in cement, iron and steel and chemical industries, or in biomass production where it could even amount to negative emissions (because CO_2 would be yielded twice: when taken out of the atmosphere by plants and when buried later on). If developing and emerging countries delay their participation in curbing CO_2 emissions and if the climate target is overshot due to late abatement, CCS is necessary as a mitigation measure to keep the carbon prices from skyrocketing (Krey and Riahi 2009, S96). In any case, costs would still be high for former free riders (those who benefited from but did not pay for mitigation, and for countries that could not afford to contribute to the internalisation of costs) (ibid., S106). Various technologies and their —rising—costs are given in Fig. 2.4.

Es ken nit wern zen, wen ejns is nitó.
It cannot be ten if one is missing.

Yidishe shprikhverter un redensartn. Frankfurt aM, 1908

Usually, the cost breakdown for CCS is as follows: the majority (75%) of costs are directed towards capture technology, whereas only some 25% go to transportation, injection and monitoring (Plasynski et al. 2009). The IEA, for example, predicted CCS to cost just USD 35–60 per tonne of emissions reductions by 2030. McKinsey (2009) foresaw a price of EUR 30–45 when the technology would be mature, sometime after 2030. At any rate, CCS has to be competitive with other technologies; costs must be at least as high as for fossil-fuelled power plants without CO_2 separation. As to storage costs alone, Friedmann and colleagues (2006) assumed that only a small fraction of costs would go towards accurate geological characterisation. They estimated these one-time costs to be on the order of USD 0.1/t CO_2 as the costs are spread over millions of tons that are likely to be injected into a field over a period of several decades. McKinsey (2009) assumed EUR 4–12/t for early commercial plants. In the light of the experience of the nuclear community and more recent projections in Fig. 2.4 (e.g., 100–190 USD/t for cement CCS), this optimistically low share may be questioned because **research expenditure** has risen with greater knowledge and insight into the complex proof of long-term safety. This is despite the fact that specific costs remain low (since comparatively little waste accrues per kWh produced) and, unlike in the CCS community, financing is secured due to the polluter-pays principle (approximately one EURc/kWh is charged to every electricity customer). Bonds and trust funds have been suggested as means of overcoming financial impasses in CCS (Gerard and Wilson 2009; Peña and Rubin 2008, respectively). Monitoring costs will depend on the time frame and specific quality requirements set by regulations. During the stage when CCS technology has not yet matured, R&D expenses accrue as economies of scale have not yet been realised, and the lack of experience may explain the lower quality of cost estimates. Furthermore, social costs also accumulate, such as licensing procedures, decision uncertainties and delays.

This section has thus far discussed cost exclusively in terms of the specific cost per kWh or per t of CO_2. In view of criterion #3, however, it is also necessary to consider the extraordinary cost for a national or regional economy, for example, the cost of a full-fledged operational CCS system as a fraction of the gross domestic product, GDP. The costs of most environmental measures as a fraction of GDP are usually well below, even orders of magnitude below, one-thousandth of a percent of GDP. This is presumably not the case for CCS, which might cost in the order of one percent of GDP if it were based on figures given in Fig. 2.2.

The order of magnitude can be shown using a rough estimate: Fossil fuel resources usually correspond to a small percentage, for example, 5%, of GDP. If calculated correctly, CCS may cost 50% of this share. Assuming that 20% of the fossil fuel stream ends up in the CCS process means that the CCS would cost 0.5% of GDP. Delaying the implementation of CCS for only 2 years would enable a country to "save" 1% of its GDP. This simple calculation demonstrates the potentially huge economic impact of the deployment or non-deployment of CCS on nations' economies and their global competitiveness.

2.6 Criterion #5: Implementation

Since the mid-1990s, CO_2 has been captured in the Sleipner field, which is StatoilHydro's, since 2018 Equinor ASG, natural gas reserve in the North Sea. The CO_2 is then injected into the saline aquifer of the so-called Utsira sands, 1 km below the seabed and 3000 m away from the reserve. The CO_2 needs to be separated in order to lower the natural concentration of 9% to the safe 2.5% of Europe's highest yielding natural gas reserve. The CO_2 would have been blown off had the Norwegian government not introduced a CO_2 tax of 25 EUR/t in 1991.

To date, no incidents have occurred in approximately 25 CCS (pioneering) projects worldwide. There has been an enormous upturn in the number of projects, experiments and industrial activities being launched. Half a dozen demonstration plants are in operation and dozens more are planned (Beck 2009; IEA 2009a). Despite this, large-scale deployment of CCS requires a satisfactory regulatory regime (Mace et al. 2007) and projects by major players such as Equinor, Shell and BP have been shelved due to insecure financing, lacking public subsidies and inadequate tax incentives and carbon trading. Recent assessments by the IEA confirm this view (IEA 2021).

Wos lejnger a blinder lebt, alz mer set er.
The longer a blind person lives, the more he or she sees.

Yidishe shprikhverter un redensartn. Frankfurt aM, 1908

From a broader perspective, to varying degrees **carbon "lock-in"** (Unruh 2000) is a feature of energy systems in many countries today: "fossil fuel-intensive systems perpetuate, delay or prevent the transition to low-carbon alternatives" (Sato et al. 2021). As coal is the highest CO_2-emitting energy carrier, this source is the focus here. Worldwide use of coal has actually risen (IEA 2009b), with the USA and China responsible for over 40% of the world's

CO_2 emissions (IEA 2009b, 11). In 2018, coal-fired electricity generation accounted for 30% of global CO_2 emissions (IEA 2019), half of which were from Chinese power plants (Friedlingstein et al. 2019). Nearly all emissions from Chinese power generation originate from coal (IEA 2009b, 21), and more than one-third of the US's CO_2 emissions are due to coal-fired power plants (US EIA 2011). The largest overall contributions to global CO_2 emissions were from China (28%), USA (15), EU-28 (9) and India (7) (Friedlingstein et al. 2019). This means that a significant share of the power infrastructure is tied to a conventionally high-emission path, which is systemically determined (centralised and cost-intensive) and associated with vested interests. Furthermore, in both China and USA, coal is one of the major domestic energy sources and is considered a key to their energy security. This is a powerful reason for continuing to use coal and invest in R&D (Dooley et al. 2009; Pew Center 2007, CSLF for USA, Liang et al. 2009; Dapeng and Weiwei 2009; Liu and Sims Gallagher 2010 for China). Other countries have followed the same reasoning, such as Germany (Praetorius and Schumacher 2009), the European Union as a whole (EU 2004) and India (Shackley and Verma 2008; Garg and Shukla 2009). According to the IEA, one-third of all coal-fired power plants worldwide that are not suitable for CCS will need to close before the end of their technical life (IEA 2008). There are, apparently, many sources of system inertia.

2.7 Criterion #6: Societal Issues

Apart from regulatory aspects, CCS system decisions involve many societal issues. After all, CCS long-term management reflects fundamental distributional risk-benefit asymmetry, just as in the case of radioactive waste (after Flüeler 2005b, 1):

- Local cost and risk versus general benefit (intra-generational issue);
- Lay persons' versus experts' perspectives and knowledge (evidentiary issue; that is, related to substantive and technical arguments);
- Current generations' benefits and decision-making power versus that of future generations (intergenerational issue).

The crucial point is probably the long-term dimension. Archer addressed this concisely in comparing the two waste systems, focusing on public perspectives: "One could sensibly argue that public discussion should focus on a time frame within which we live our lives, rather than concern ourselves with climate impacts tens of thousands of years in the future. On the other hand, the 10 kyr lifetime of nuclear waste seems quite relevant to public perception of nuclear energy decisions today" (Archer 2005, 5). Ha-Duong and Loisel reached the following conclusion from their stakeholder analysis: "Zero is the only acceptable leakage rate for geologically stored CO_2" (Ha-Duong and Loisel 2009). Various studies have analysed the so-called "**acceptance**" of CCS. A common denominator of these studies is the rather poor state of knowledge and the need for "outreach" (Plasynski et al. 2009), following the "**deficit model**" (Sect. 6.2, Question 3 in Chap. 6). This means that specialists only have to inform non-specialists at large to overcome the "information gap" between them and the public. The strenuous experience in radioactive waste governance shows that this cannot be enough. The decision-making process from a generic approval of CCS to site selection, risk assessment, monitoring, accounting and closure must be transparent, open, fair and accountable, with well-defined decisive criteria in order for CCS to contribute to a sustainable climate policy. Recent "best practice" guides subsume demographic trends and land use under the heading "social context analysis" (NETL 2017). According to researchers closely following the field, public participation is still instrumentally motivated by project proponents (Reins 2017).

Provided that the risks of sequestration can be minimised (especially earthquakes and contamination of drinking water aquifers through leakage), public acceptance of CCS may be higher than for radioactive waste as the safety periods are very different: for CO_2 in the order of 10,000 years, for high-level radioactive waste in the order of 1 million years.

Der bester schuster fun ale schnajders is Yankl der beker.
The best shoemaker of all tailors is Yankel the baker.

Yidishe shprikhverter un redensartn. Frankfurt aM, 1908

2.8 Some Conclusions

Since innovation not only consists of technological progress but also is an integration of technical, conceptual, organisational and societal processes, a one-sided perspective is prone to failure. The following aspects are crucial to the CCS innovation process (and potentially others):

- Integrate the notion of time: needed period of isolation, safety assessment, processes (geological formations, mineralisation) and implementation of the CCS system;

- Consider adequate integration: integrate scales (geographical, temporal, institutional) into overall energy system(s), markets and regulation;
- Consider CO_2 disposal as a goal in itself, not a "by-product" of other economic activities;
- Optimise CCS to environmental performance (with regard to effectiveness, efficiency and timing, for example, new "capture-ready" coal plants to be commissioned on the premises that old ones are shut down);
- Settle major issues (above) and harmonise regulation before large actors (states and companies) deploy CCS on a scale of a *fait accompli* (technically and in terms of generating accountable CO_2 certificates);
- Employ social sciences and humanities, recognising that they can contribute more than just assessing the technology costs and increasing public acceptance (Spreng et al. 2012);
- Before CCS can be considered a viable and sustainable option, rules and institutions for an international, reliably working CCS regime must be developed. Suitable mechanisms for enforcing regulation are part of any plausible regime; the global financial crisis of until 2009 may serve as an instructive case. Perhaps, it is necessary to have an "entrance exam" for countries wanting to use CCS as an emissions reduction measure, similar to the scrutiny subjected to applicants for admission to the eurozone. Funds for CCS may have to be deposited in an internationally monitored fund before CCS can start;
- Take action before political pressures and/or environmental failures (such as spillouts) make it compulsory.

Questions

1. Comment on the proposition: "technology changes exponentially, but social, economic and legal systems change incrementally".
2. What is meant by "lock-in" in the context of energy policy?
3. What is the short definition of "Recasting all complex social situations either as neat problems with definite, computable solutions or as transparent and self-evident processes that can be easily optimised—if only the right algorithms are in place"?

Answers

1. This is how Larry Downes formulated the so-called "pacing problem", meaning that technological innovation outpaces the ability of laws and regulations to keep up (Downes 2009).

2. A society is caught in a technology or technologies based on energy carriers, e.g., fossil fuels. A lock-in is a type of path dependency where decisions depend on prior decisions or past activities, foreclosing alternative ways (oil-fired heating keeping house owners from investing in renewables, spark-ignited cars running on petrol standing in the way of e-vehicles, etc.).

3. Technical fix. Quote from Morozov (2013). And the author goes on: "… this quest is likely to have unexpected consequences that could eventually cause more damage than the solutions they seek to address" (ibid., 5).

Exercise

What would an "ideal" harmonised regulation and control system look like? What would be the incentives and benefits for everyone to participate?

Additional Information

All weblinks accessed 27 January 2023.

Key Readings

Key Reference

Flüeler T (2006b) etc.: other own references in Annex
Flüeler T (2012d) Technical fixes under surveillance—CCS and lessons learned from the governance of long-term radioactive waste management (Chap. 10). In: Spreng D, Flüeler T, Goldblatt D, Minsch J (eds) Tackling long-term global energy problems: the contribution of social science. Environment & Policy, vol 52. Springer, Dordrecht NL, pp 191–226. https://doi.org/10.1007/978-94-007-2333-7

Problem Perception

EC, European Commission, European Parliament (Jan–Feb 2009) Europeans' attitudes towards climate change. Special Eurobarometer 313/Wave 71.1. Report. Directorate-General for Communication, Brussels
EU, European Union (2019) Special Eurobarometer 490. Report. Climate change. https://data.europa.eu/data/datasets/s2212_91_3_490_eng?locale=en
Pew Research Center (2019) Climate change still seen as the top global threat, but cyberattacks a rising concern, 35 pp. https://www.pewresearch.org/global/2019/02/10/climate-change-still-seen-as-the-top-global-threat-but-cyberattacks-a-rising-concern/

Technology Policy, Technology Assessment

Kasperson RE, Golding D, Tuler S (1992) Social distrust as a factor in siting hazardous facilities and communicating risks. Social Issues 48(4):161–187. https://doi.org/10.1111/j.1540-4560.1992.tb01950.x

Morone JG, Woodhouse EJ (1986) Averting catastrophe. Strategies for regulating risky technologies. University of California Press, Berkeley

Ravetz JR (1980) Public perceptions of acceptable risks as evidence for their cognitive, technical and social structure. In: Dierkes M, Edward S, Coppock R (eds) Technological risk. Its perception and handling in the European community. Oelgeschlager, Gunn and Hain, Königstein, Germany, pp 45–54

Ropohl G (1999) Philosophy of socio-technical systems. Soc Philos Technol Q Electron J 4(3):186–194. https://doi.org/10.5840/techne19994311

Spreng D, Flüeler T, Goldblatt DL, Minsch J (eds) (2012) Tackling long-term global energy problems: the contribution of social science. Environment & Policy, vol 52. Springer, Dordrecht NL

Vlek C, Stallén PJ (1981) Judging risks and benefits in the small and in the large. Organ Behav Hum Perform 28:235–271

Wynne B (1980) Technology, risk, and participation: the social treatment of uncertainty. In: Conrad J (ed) Society, technology, and risk assessment. Academic Press, London, pp 83–107

Technical Fixes, Etc.

Metlay D (1978) History and interpretation of radioactive waste management in the United States. In: Bishop WP, Hoos IR, Hilberry N, Metlay DS, Watson RA (eds) Essays on issues relevant to the regulation of radioactive waste management. PB–281347. NUREG-0412. Sandia Labs, Albuquerque, NM. United States Nuclear Regulatory Commission, NRC, Washington, DC, pp 1–19

Morozov E (2013) To save everything, click here. The folly of technological solutionism. PublicAffairs, New York

Oreskes N, Conway EM (2010) Merchants of doubt. How a handful of scientists obscured the truth on issues from tobacco smoke to global warming. Bloomsbury Press, New York

Pielke R Jr (2010) The climate fix: what scientists and politicians won't tell you about global warming. Basic Books, New York

Science and Technology Studies

Bijker WE, Hughes TP, Pinch T (1987) The social construction of technological systems. New directions in the sociology and history of technology. MIT Press, Cambridge, MA

Collins HM (1981) Stages in the empirical programme of relativism. Soc Stud Sci 11:3–10

Latour B (1987) Science in action. How to follow scientists and engineers through society, Harvard University Press, Cambridge, MA

Nuclear Waste (Extended in Chap. 5)

ARGONA (2017) Analysis for risk governance. EU project. https://cordis.europa.eu/project/id/36413/de

Blowers A, Lowry D, Solomon BD (1991) The international politics of nuclear waste. St. Martin's Press, New York

Carter LJ (1987) Nuclear imperatives and public trust: dealing with radioactive waste. Resources of the Future, Washington, DC

Colglazier EW (1982) The politics of nuclear waste. Pergamon New York

COWAM (2003) COWAM network. Nuclear waste management from a local perspective. Reflections for a better governance. https://cordis.europa.eu/project/id/FIKW-CT-2000-20072/reporting

Freudenburg WR (2004) Can we learn from failure? Examining US experiences with nuclear repository siting. J Risk Res 7(2):153–169

IAEA, International Atomic Energy Agency (1998) Technical, institutional and economic factors important for developing a multinational radioactive waste repository. TECDOC-1021, IAEA, Vienna. https://www.iaea.org/publications

IAEA (2004) Developing multinational radioactive waste repositories: infrastructural framework and scenarios of cooperation. TECDOC-1413. IAEA, Vienna

Ilg P, Gabbert S, Weikard HP (2017) Nuclear waste management under approaching disaster: a comparison of decommissioning strategies for the German repository Asse II. Risk Anal 37(7):1213–1232. https://doi.org/10.1111/risa.12648

Kasperson RE (ed) (1983) Equity issues in radioactive waste management. Oelgeschlager, Gunn and Hain, Cambridge

NEA, Nuclear Energy Agency (2000) Nuclear energy in a sustainable development perspective. OECD, Paris. https://www.oecd-nea.org/jcms/pl_13434/nuclear-energy-in-a-sustainable-development-perspective?details=true

North DW (1999) A perspective on nuclear waste. Risk Anal 19(4):751–758, 751. https://doi.org/10.1111/j.1539-6924.1999.tb00444.x

Solomon BD, Andrén M, Strandberg U (2010) Three decades of social science research on high-level nuclear waste: achievements and future challenges. Risk, Hazards Crisis Public Policy 1:13–47. https://doi.org/10.2202/1944-4079.1036

Carbon Capture and Storage, CCS

Archer D (2005) Fate of fossil fuel CO_2 in geologic time. J Geophys Res 110(C09S05):1–6

Beck B (Jan 2009) The technology and status of carbon dioxide capture and storage. IEA greenhouse gas R&D programme. Presentation. City University, London

Benson SM (2005) Lessons learned from industrial and natural analogs for health, safety and environmental risk assessment for geologic storage of carbon dioxide. In: Thomas DC, Benson SM (eds) Carbon dioxide capture for storage in deep geologic formations—results from the CO_2 capture project, vol 2. Elsevier, Amsterdam, pp 1133–1141

CSLF, Carbon Sequestration Leadership Forum (established in June 2003) https://www.cslforum.org/cslf/Projects/Summaries

Dapeng L, Weiwei W (2009) Barriers and incentives of CCS deployment in China: results from semi-structured interviews. Energy Policy 37:2421–2432

de Coninck H (2008) Trojan horse or horn of plenty? Reflections on allowing CCS in the CDM. Energy Policy 36:929–936

Dixon T (2009) International legal and regulatory developments for carbon dioxide capture and storage: from the London Convention to the Clean Development Mechanism. Proc Inst Mech Eng, Part A: J Power Energy 223(3):293–297

Dooley JJ, Dahowski RT, Davidson CL (2009) The potential for increased atmospheric CO_2 emissions and accelerated consumption of deep geologic CO_2 storage resources resulting from the large-scale deployment of a CCS-enabled unconventional fossil fuels industry in the US. Int J Greenhouse Gas Control 3(6):720–730

EU CO_2 Network, European Carbon Dioxide Network (2004) Capturing and storing carbon dioxide: technical lessons learned. http://www.co2geonet.com/home/, https://ec.europa.eu/clima/eu-action/carbon-capture-use-and-storage_en

Feroz EH, Raab RL, Ulleberg GT, Alsharif K (2009) Global warming and environmental production efficiency ranking of the Kyoto Protocol nations. J Environ Manage 90(2):1178–1183

Friedlingstein P et al (2019) GCB, global carbon budget. Earth Syst Sci Data 11:1783–1838. https://doi.org/10.5194/essd-11-1783-2019

Friedmann SJ, Dooley JJ, Held H, Edenhofer O (2006) The low cost of geological assessment for underground CO_2 storage: policy and economic implications. Energy Convers Manage 47:1894–1901

Gale J, Read T (2005) Rules and standards for CO_2 capture and storage: considering the options. In: Wilson M et al (eds) Greenhouse gas control technologies, vol 2. Elsevier, Amsterdam, pp 1461–1466

Garg A, Shukla PR (2009) Coal and energy security for India: role of carbon dioxide (CO_2) capture and storage (CCS). Energy 34(8):1032–1041

Gerard D, Wilson EJ (2009) Environmental bonds and the challenge of long-term carbon sequestration. J Environ Manage 90(2):1097–1105

Grünwald R (2008) Treibhausgas – ab in die Versenkung? Möglichkeiten und Risiken der Abscheidung und Lagerung von CO_2. Studien des Büros für Technikfolgen-Abschätzung beim Deutschen Bundestag. Global zukunftsfähige Entwicklung, Bd 25. Edition Sigma, Berlin

Ha-Duong M, Loisel R (2009) Zero is the only acceptable leakage rate for geologically stored CO_2: an editorial comment. Clim Change 93 (3–4):311–317

Haefeli S, Bosi M, Philibert C (2005) Important accounting issues for carbon dioxide capture and storage projects under the UNFCCC. In: Rubin ES, Keith DW, Gilboy CF (eds) Greenhouse gas control technologies, vol I. Elsevier, Amsterdam, pp 953–960

IEA, International Energy Agency (2008) Energy technology perspectives 2008. Scenarios and strategies to 2050. Executive summary. OECD/IEA, Paris

IEA (8 July 2009a) Carbon capture and storage. Full-scale demonstration progress update. OECD/IEA, Paris

IEA (2009b) IEA statistics. CO_2 emissions from fuel combustion. Highlights. Paris, OECD/IEA

IEA (2009c) Technology roadmap. Carbon capture and storage. OECD/IEA, Paris

IEA (2019) Global energy and CO_2 status report 2019. OECD/IEA, Paris. https://www.iea.org/reports/global-energy-co2-status-report-2019/emissions

IEA (2021) The role of CCUS in low-carbon power systems. https://www.iea.org/reports/the-role-of-ccus-in-low-carbon-power-systems

IPCC, Intergovernmental Panel on Climate Change (2005) IPCC special report on carbon dioxide capture and storage. In: Metz B, Davidson O, de Coninck HC, Loos M, Meyer LA (eds) Prepared by working group III. Cambridge University Press, Cambridge

IRGC, International Risk Governance Council (2007) Workshop on regulation of carbon capture and storage. March 15 and 16. IRGC, Geneva

IRGC (2008) Regulation of carbon capture and storage. Policy Brief. IRGC, Geneva

Koornneef J, Faaij A, Turkenburg W (2008) The screening and scoping of environmental impact assessment and strategic environmental assessment of carbon capture and storage in the Netherlands. Environ Impact Assess Rev 28:392–414

Krey V, Riahi K (2009) Implications of delayed participation and technology failure for the feasibility, costs, and likelihood of staying below temperature targets—greenhouse gas mitigation scenarios for the 21st century. Energy economics 31, Suppl. 2, International, U.S. and E.U. climate change control scenarios: Results from EMF 22, December 2009, S94-S106

Liang X, Reiner D, Gibbins J, Li J (2009) Assessing the value of CO_2 capture ready in new-build pulverised coal-fired power plants in China. Int J Greenhouse Gas Control 3(6):787–792

Liu H, Sims Gallagher K (2010) Catalyzing strategic transformation to a low-carbon economy: a CCS roadmap for China. Energy Policy 38(1):59–74

Mace MJ, Hendricks C, Coenraads R (2007) Regulatory challenges to the implementation of carbon capture and geological storage within the European Union under EU and international law. Int J Greenhouse Gas Control 1:253–260

NETL, National Energy Technology Laboratory (2017) Best practices: site screening, site selection, and site characterization for geologic storage projects. 2017 revised edition. DOE/NETL-2017/1844. NETL, Albany, OR. https://www.netl.doe.gov/

Page SC, Williamson AG, Mason IG (2009) Carbon capture and storage: fundamental thermodynamics and current technology. Energy Policy 37(9):3314–3324

Peña N, Rubin ES (2008) A trust fund approach to accelerating deployment of CCS: options and considerations. Coal initiative reports. White paper series. Pew Center, Arlington, VA

Pew Center (2007) A program to accelerate the deployment of CO_2 capture and storage (CCS): rationale, objectives, and costs. Coal initiative reports. White paper series. Pew Center, Arlington, VA

Plasynski SI, Litynski JT, McIlvried HG, Srivastava RD (2009) Progress and new developments in carbon capture and storage. Crit Rev Plant Sci 28(3):123–138

Pollak M, Wilson EJ (2009) Risk governance for geological storage of CO_2 under the Clean Development Mechanism. Clim Policy 9(1):71–87

Powicki CR (2007) Closing the carbon cycle. EPRI J. Spring 2007. Electric Power Research Institute (EPRI), Palo Alto, CA

Praetorius B, Schumacher K (2009) Greenhouse gas mitigation in a carbon constrained world: the role of carbon capture and storage. Energy Policy 37(12):5081–5093

Reins L (2017) Regulating shale gas. The challenge of coherent environmental and energy regulation. Leuven Global Governance series. Edward Elgar, Cheltenham Glos, UK

Sato, I, Elliott B, Schumer C (2021) What is carbon lock-in and how can we avoid it? World Resources Institute, Washington, DC. https://www.wri.org/insights/carbon-lock-in-definition

Shackley S, Verma P (2008) Tackling CO_2 reduction in India through use of CO_2 capture and storage (CCS). Prospects and challenges. Energy Policy 36:3554–3561

Steeneveldt R, Berger B, Torp TA (2006) CO_2 capture and storage: closing the knowing doing gap. Chem Eng Res Des 84(9):739–763

Stern N (2007) The economics of climate change. Cambridge University Press, Cambridge

Unruh GC (2000) Understanding carbon lock-in. Energy Policy 28:817–830

US EIA, Energy Information Administration (2011) Emissions of greenhouse gases report. https://www.eia.gov/environment/emissions/ghg_report/

WRI, World Resource Institute (2008) CCS guidelines. Guidelines for carbon dioxide capture, transport and storage. WRI, Washington, DC

Wuppertal Institute, DLR, ZSW, PIK (2008) RECCS. Ecological, economic and structural comparison of renewable energy technologies (RE) with carbon capture and storage (CCS). An integrated approach. Published by the Federal Ministry for the Environment, Nature Conservation and Nuclear Safety BMU. Wuppertal Institute, Wuppertal

Websites

https://www.iea.org/ (International Energy Agency: general, CCS)

https://www.oecd-nea.org/ (Nuclear Energy Agency of the OECD: radioactive waste)

https://www.ipcc.ch/ (Intergovernmental Panel on Climate Change: carbon storage)

https://www.globalccsinstitute.com/ (Global CCS Institute: carbon capture and storage), datablase: https://co2re.co/

Sources

Downes L (2009) The laws of disruption: harnessing the new forces that govern life and business in the digital age. Basic Books, New York (question 1)

Europäische Akademie (Decker M, Ladikas M) (2004) Technology assessment in Europe. Between method and impact. Springer. https://doi.org/10.1007/978-3-662-06171-8_1, p 4 (Box 2.1)

Geneva Environment Network (2022) Update (17 Oct). Climate-altering technologies and measures. https://www.genevaenvironment network.org/resources/updates/climate-altering-technologies-and-measures

Goldman Sachs (2019) Carbonomics. The future of energy in the age of climate change. Equity research. 11 Dec 2019, pp 14–15 (Figs. 2.2 and 2.4)

Grunwald A (2010) Technikfolgenabschätzung – eine Einführung. edition sigma, Berlin (Box 2.1)

IPCC (2013) Summary for policymakers. In: Climate change 2013: the physical science basis. In: Stocker T et al (eds) Contribution of working group I to the Fifth assessment report of the Intergovernmental Panel on Climate Change. Cambridge University Press, Cambridge, UK, p 29 (Box 2.2)

Ki-moon B (2019) Governing geoengineering. Project Syndicate. 11 March. https://www.project-syndicate.org/commentary/climate-change-geoengineering-technologies-governance-by-ban-ki-moon-2019-03, https://environmentassembly.unenvironment.org/unea4 (quote Sect. 2.3)

McKinsey (2009) Pathways to a low-carbon economy. Version 2 of the global greenhouse has abatement cost curve. McKinsey and Company. http://www.mckinsey.com (use search function) (Fig. 2.3)

Yidishe shprikhverter un redensartn (1908, 1949, 1965/1984) Bernstein I, Segel BW (eds) Frankfurt aM, 1908 (Yiddish proverbs. Schocken Books, New York, 1949; Jiddische Sprichwörter. Insel-Bücherei Nr. 828. Insel-Verlag, Bern, 1965/1984) (transl. to English tf)

Goals—Needs, Rules and Procedures

3

Abstract

Following the framework of Chap. 2, the necessity to deploy a programme (in waste management) has to be shown. Section 3.3 is on goals, in a novel and productive way in the sense that the buzzword "sustainable development" (Sect. 3.1) is "deconstructed" (passive protection vs. active intervention) and areas of "consensus" and "compromise", respectively, are identified. This is useful as the three mentioned waste disposal systems indisputably are highly controversial, almost intractable, sociotechnical issues. All the more, it is pivotal to proceed in a comprehensive, transparent and participative manner. The chapter suggests fundamental rules to follow, based on an amplified notion of sustainable development, to eventually find "common ground" (with a common goal) to reach respective long-lasting "solutions".

Keywords

Multiperspectiveness • Controversies • Consensus/consent • Discourse • Problem recognition • Sustainable development • Rights of future generations • Chain of obligation • Rolling present • Procedural strategy

Learning Objectives

- Understand what multiperspectiveness in complex, "messy" problems means;
- Recognise the usefulness of the notion "sustainable development";
- Understand the logic of the three-step approach to find "common ground".

3.1 Integrated Perspective Wanted

Dealing with complex sociotechnical systems[1] such as the disposition of radioactive waste, conventional toxic waste or carbon (storage) needs an integrated perspective. Much of the widespread blockage faced in these sensitive policy areas may be ascribed to the neglect of looking at the various dimensions involved, such as technical, social and political. This multidimensionality requires an appropriate reference system. Normatively, the principle of sustainable development (incorporating protection as well as control, Box 3.1) seems to suggest itself, for two reasons. Firstly, it facilitates a stepwise analysis according to various dimensions: not only the triad of ecological, economical and social but also temporal, spatial, technical, political and ethical (Fig. 3.1). Secondly, it forces upon stakeholders, including decision-makers, an examination of these dimensions and, consequently, it is apt to incorporate all/most parties' perspectives, needs, targets/goals and knowledge systems (Table 3.1). As "good" decisions are always goal-related (good with respect to "what"), the goal discourse is vital and must be thoroughly led.

> **Box 3.1: Sustainable Development**
> "Sustainable development is development that meets the needs of the present without compromising the ability of future generations to meet their own needs." (WCED 1987, 43)[2]

---

[1] For key terms and concepts refer to Glossary in the back.

[2] Obviously, there is a more fundamental nexus of sustainable development and waste as current societies build on innovation which, by itself and with its "creative destruction" (Schumpeter 1942; Sombart 1916), is connoted to a devaluation of older items and strategies (cf. Grunwald 2016, 326).

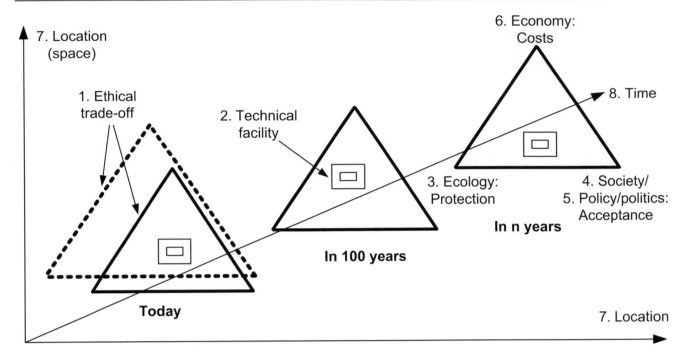

Fig. 3.1 Sustainability of disposition systems. Eight dimensions—not just the three classical ones of the "magical" triangle Ecology-Society-Economy—have to be considered: an ethical trade-off takes place in the design of the facility (technical dimension), along the ecological dimension (protection of humans and the environment), the social and political dimensions (society and balance of power determine acceptance) as well as the economical dimension (costs of disposition including institutional control). This decision bears an eminent spatial (location) and temporal dimension (period of isolation and concern) (after Flüeler 2001a/2006b, 202)

Table 3.1 Where to possibly reach "common ground" and where not. Relations (and hierarchy) of consent and dissent at diverse levels and with diverse conflict types. The middle region is amenable to good chances for some "common ground", being above the non-negotiable core beliefs and below practical project management where a compromise is best to achieve (from Flüeler 2006b, 214, expanded)

Level	Conflict type	State of agreement	Perspective/goal/fields (examples)
Secondary beliefs			
Implementation (dependent on policies, funding, authority)	Judgement/interest/relational conflict	Compromise	"Real" project/site
Procedure/methodology	Judgement/target/interest conflict	Consensus	Siting, monitoring
Roles, decisions (instrumental and institutional goals)	Judgement/target/distributional/interest conflict		Performance assessment, quality assurance, inclusive reviewing
Protection goals (passive protection, active control, involvement, power of decision) (= "success criteria")	Value/target/distributional/judgement conflict	Consensus	Safety and control goals
Factual constraints	Judgement/interest conflict	Consensus	Waste existent
Concept of sustainability	Value/target conflict	Compromise ("weak" sustainability[a])	Practical trade-off of dimensions (technical and social goals)
Core beliefs			
Attitudes of stakeholders	Value/target conflict	Dissent	Pro- *versus* anti-nuclear
Models of rationality	Value/target conflict	Dissent	Technocentric/anthropocentric *versus* biocentric or even ecocentric worldview

[a] "Weak" sustainability allows for substantial substitutability of resources (Solow 1974)

The long-term objective (ecological) dimension of highly toxic waste is of outstanding ethical relevance. The ones who make the profit (e.g., of energy of which waste is a result) most likely do not bear—possible—risks from the waste (Fig. 3.2). The decision situation is such that the current generations (we!) have to decide (postponement is also a decision), and apart from winners (this waste-producing society), there will be losers (locals at a chosen site and future generations). This is a formidable risk-benefit asymmetry. The aim is to obtain what Majone (1989) (and others) termed a "multiple evaluation", acknowledging the legitimacy of different criteria and perspectives as being more than the sum of partial appraisals. To be able and competent to handle such complex issues requires adequate tools: a useful procedure (below) and a good technical analysis (Chap. 4) to set up a targeted programme (Chap. 5) followed by its monitoring (Chap. 6). The basic normative requirement would be the following: today's acting generations may—or should—begin the long-term programmes if their actions and consequences thereof do not violate today's norms and conceivable ones in the future. Future actors should not be obliged to violate present and their own, future, norms.

To address the rights and needs of future generations, the ethical concepts of the "chain of obligation" and the "rolling present" were introduced (Box 3.2).

Box 3.2: Chain of Obligation

"Stretches from the present into the indefinite future, and unless we ensure conditions favourable to the welfare of future generations, we wrong our existing children in the sense that they will be unable to fulfill their obligations to their children while enjoying a favourable way of life themselves … our responsibility for the distant future follows directly from our obligation to our existing children, not to undifferentiated potential beings …." (Howarth 1992, 133, 138)

Rolling Present

"According to [the] idea of a 'rolling present' the current generation would have a responsibility to provide to the next succeeding generation the skills, resources and opportunities to deal with any problem the current generation passes on. However, if the present generation delays the construction of a disposal facility to await advances in technology, or because storage is cheaper, it should not expect future generations to make a different decision. Such an approach in effect would always pass responsibility for real action to future generations and for this reason could be judged unethical." (NEA 1995, 10; see also Socolow and English 2011, 191)

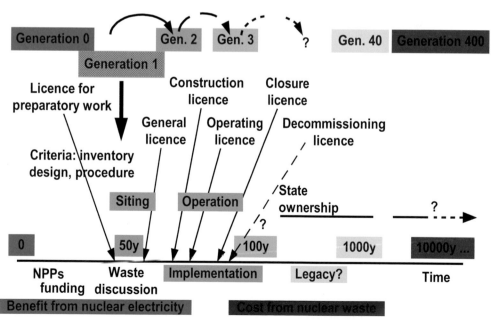

Fig. 3.2 Radioactive waste management has a long-term safety and a long-term project character. It has to be supported by the technical community, the political decision makers and the general public over decades. While still benefiting from nuclear electricity, we, at present, are "Generation 1" having to start implementing radioactive waste management. Some duties—of monitoring, etc.—will have to be handed over to "Generation 2" being at the edge of merely bearing risks from waste. At any rate, the duties of regulatory bodies stand out in long-lasting licensing (reproduced from Flüeler 2004a/Flüeler 2006b, 204)

3.2 The Search for Reasons for Fundamental Controversies

For diverse reasons, the governance of radioactive waste has always been regarded as a particularly difficult issue. Somewhat prophetically, the Swiss Government, the Federal Council, focused on it in their 1957 Statement to the Atomic Energy and Radiation Protection Article in the Federal Constitution: "… perhaps the response to the question whether the atomic ash … will be rendered harmless or may even be utilised productively will decide upon the way how to apply atomic energy in the future" (Swiss Government 1957, 1141ff).

Two decades ago, D. W. North, a long-standing expert of the scene, wrote in Risk Analysis, a journal renowned beyond the nuclear community: (High-level) "nuclear waste management has the deserved reputation as one of the most intractable policy issues facing the United States and other nations using nuclear reactors for electric power generation" (North 1999, 751). The issue is, thus, far away from being "closed" in the science and technology studies sense. The situation is so muddled that even the originally technocratic International Atomic Energy Agency, IAEA, sighted the topic, e.g., in the seminal Córdoba Conference in March 2000: "In almost all of the Conference's technical sessions, there was discussion of the need to involve all interested parties ['stakeholders'] in the decision-making processes related to radioactive waste management" (IAEA 2000, vi). This need has been evolvingly perceived virtually as a recurrent theme of the "nuclear dispute" worldwide in recent years.

According to the perspective chosen on the political and societal stage, various and diverse levels and basics become apparent on how to judge radioactive waste.

First Reflection

On a surficial level (Level 1), the current situation in the radioactive waste arena in a country for some is such that it is solely a "political", i.e., a technically solved, issue, for others it is an unsolved, even "unsolvable" problem. There is a stalemate between implementers and regulators (who have to erect and license disposal facilities, respectively) on the one side and national as well as local opposition (who raise the waste issue to the level of energy policy and/or block construction) on the other side. The actual issues are repository projects and preparatory activities formerly at Wellenberg, in Benken (both Switzerland), Bure (France), Gorleben (Germany) or Yucca Mountain (USA). Two "camps" face each other: the camp of the implementers (Nagra, US DOE)/safety authorities and the camp of (regional and national) resistance (Fig. 3.3).

It can be different, though. In Sweden, there was a competition between two municipalities, Oskarshamn and Östhammar/Forsmark, both "nuclear" communities, wanting a repository to secure working places. The implementer SKB would grant three-quarters of the "benefit package" (of around 300 million EUR) to the community which would not be awarded the contract. Finally, in 2009, they agreed on the compensation: Östhammar would get the high-level waste repository (plus taxes and working places), Oskarshamn the bulk of the package plus a factory to build the copper canisters for the spent fuel (encapsulation plant) (NEA 2012; Kojo and Richardson 2009).

Second Reflection

On a closer reflection, one gets to Level 2: The complexity of a problem is defined by the number of its elements as well as their diversity and interrelations which vary with time. Decision-making processes as in radioactive waste

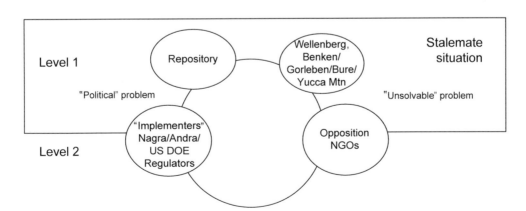

Fig. 3.3 Political stalemate. In daily politics two "camps" face each other: the "implementers" (in Switzerland: Nagra, USA: department of energy (DOE), France: ANDRA) and the regulators on the one side and the opposition on national and local levels on the other side. They were and are in a dispute over a project (repository) at an actual site (Wellenberg in the 1990s, Zurich Northeast/Benken currently, Bure in France, then Gorleben in Germany or Yucca Mountain in the USA) (reproduced from Flüeler 2006b, 10)

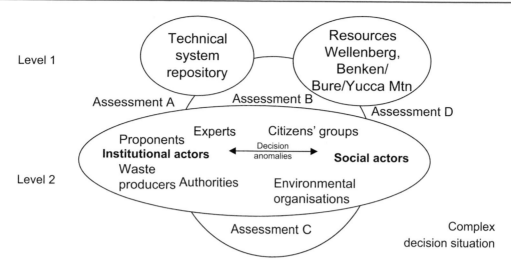

Fig. 3.4 Complex decision situation. On Level 2, a much more subtle decision situation is displayed than was presumed for Level 1. Both "camps" consist of diverse institutional and social actors who do not plainly accept or reject a project (generally: a technical system) at a given site (a resource). They carry out differentiated assessments (A through D) which might go wrong for traditional "camp (group) thinking" (reproduced from Flüeler 2006b, 11)

governance are undoubtedly complex problem situations (Fig. 3.4). According to the actor model (Rohrmann 1991; Wiedemann 1991), the implementing organisations (Nagra, once GNW), as well as the authorities, the environmental organisations, and some experts are institutional actors; citizens' groups and other experts are non-institutionalised, social actors. Waste producers are institutional (NPP operators, indirectly Office of Public Health by collecting, e.g., waste from hospitals) and social actors (electricity consumers). The actors usually have ambi- or even polyvalent relations; correspondingly, the judgements are not plainly pro or con, but there are different views on a technical project or a resource, respectively (assessments A, B …). The consequences thereof are individual and institutional decision anomalies (Klose 1994, 138ff), meaning—in simplified terms—ways to behave which cannot be followed by other actors (e.g., regarding the principle of sustainability).

Third Reflection

The base for decision anomalies (also for the stalemate situation in the first reflection) is the fact that the actors start out from diverse reality models whose grounds in turn are diverse rationality concepts. Such interrelations become apparent only if Level 3, the level of "mental models" (Jones et al. 2011), with the "basic underlying assumptions" according to Schein (1985), is considered (Fig. 3.5). These assumptions make understandable how the technical systems and their safety/risk analyses are erected and how they are assessed from diverse viewpoints, well beyond stakeholders' preferences. With respect to science, a convergent approach was chosen by Kuhn (1962ff) in analysing the social influence on scientific knowledge when he coined the term "paradigm". In this context, its sociological sense is pertinent, viz., it is "the entire constellation of beliefs, values, techniques, and so on shared by the members of a given [scientific] community" (Kuhn 1996, 175). These considerations have a bearing on how to evaluate the uncertainty notion, as well as expert and value judgments and paramount aspects, in the issues under study.

The multidimensionality of the issue suggests the application of the principle of sustainability or sustainable development (Box 3.1). Therefore, it is based on two pillars: protection, i.e., passive safety, and intervention potential, i.e., the ability of today's and future generations to control. The decision on that is both scientific-technical and societal, amounting not only to a "co-production", i.e., "the simultaneous production of knowledge and social order", as the interplay of various spheres in society may be termed (Jasanoff 1995, 393), but also possibly to a "co-decision" on radioactive waste, conventional toxic waste or carbon storage.

As we deal with complex systems, a high degree of division of labour as well as division of knowledge (Rammert 2016), we also need to integrate this knowledge if we want to reach a co-production and, in the end, a co-decision on the contentious policy issues under scrutiny. This presupposes an approach of multiperspectiveness and of integrative discourse.

Fig. 3.5 Mental models. Only Level 3 revealing the mental models of the actors (for their concepts of rationality, reality or risk) makes it comprehensible that the concerned parties have particular perspectives resulting in different analysis, assessment and decision making with regard to technology, resources and corresponding risks. Therefore, the "acceptance" of a risk, e.g., due to the disposal of radioactive or conventional waste or CO_2, is not given. Significant divergences lead to a disturbed (risk) communication (reproduced from Flüeler 2006b, 12)

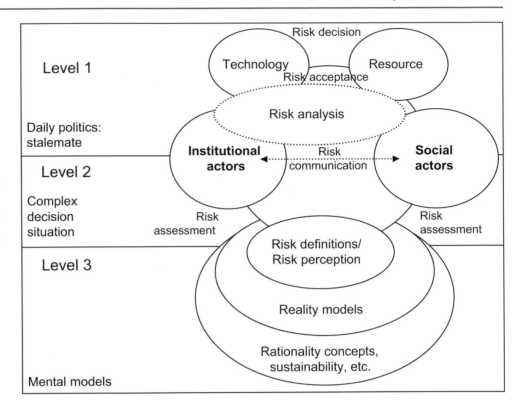

3.3 The Search for "Common Ground"

In view of a common understanding to reach a "solution" and a successful respective transfer of knowledge, it is vital to explore contextual issues and tacit/implicit knowledge—they determine the degree of societal understanding of the eventual disposition system. Unless the rationale of conceptual reasoning is appropriately handed over to the next—technical, political and societal—generations, the entire undertaking is bound to fail (Flüeler 2005c).

Back in the 1980s, Parker and colleagues long-sightedly concluded that, in the end, one can only "… try to develop as broad a consensus as possible in support of the solution that is finally reached. It has to be recognized that there will always be an irreducible amount of uncertainty in the outcome of any solution" (Parker et al. 1984, 101). Over 30 years ago, Carter (1987, 427) called in to find a "common ground" in managing radioactive waste yet without specifying. By focusing on "common ground", rather than "consensus", it has to be emphasised that it is not only intended to call for as many voices but also for as many perspectives as possible so as to incorporate all relevant facets in the dimensional discourse: ethical, technical, ecological, economical, political, societal, spatial and temporal. Consensus, at that, would probably amount to majoritarian deliberation anyhow (Herzig and Statham 1993), i.e., the rule of the majority. This is not to avoid the issue of representativeness or, by no means, to devitalise claims for

wider participation (e.g., OECD 2020), but to focus on a dimensional discourse as inclusively as possible. In view of this multidimensionality, it is also an avenue to find society's way to some sort of sustainable "closure" of the issue (Berkhout 1991). On this background, it is useful to specify what might be understood by "common ground". Trying to decompose ever-used buzzwords like "consensus" or "compromise" one may outline where and how "common ground" is likely to be achieved (Table 3.1).[3] It cannot be assumed to reach consensus "at heart", in the stakeholders' core beliefs (Sabatier 1987). Society must, however, agree on three levels:

- Problem recognition: The waste exists, the problem must be tackled, eventually "solved", at least set on track to be solved;
- Main goal consensus: The degree of protection and intervention must be defined; according to the scientific consent, passive safety must prevail;
- Procedural strategy: The "rules of the game" (to find a suitable site and to implement disposal) have to be clear from the outset.

The involvement of stakeholders requires consideration of as many relevant perspectives (not necessarily as many

---

[3] See Glossary for differences between "compromise", "consensus" and "consent".

individuals as possible). Pros and cons must be thoroughly scrutinised, to successfully "close", i.e., resolve, certain issues and proceed to the following step, stage or phase, eventually and ideally to reach a "co-decision". In conflicts, it is productive to focus on interests rather than positions, as the problems are determined by interests; these are wider than positions and the chance to agree upon them is greater (Fisher et al. 1981/1991). The long-time process has to be overseen which is covered in Chaps. 5, 6 and 7.

3.4 Three-Step Approach

The nuclear community recognises that the long-term safety of repositories "is not … a rigorous proof of safety … but rather a convincing set of arguments" (NEA 1999). It has, however, been difficult to "live" its sociotechnical nature (NEA 2013). Although the waste problem is driven by technology and, indeed, a technological constraint, in the end, it has to be solved by society. Building upon the defence-in-depth principle, the concept of integral, technical and societal, robustness was developed (Flüeler 2002d/ 2006b) and is reproduced in Chaps. 4 and 5. A system is *socially robust* if most arguments, evidence, social alignments, interests and cultural values lead to a consistent option (Rip 1987). The concept attempts to consider technical and social issues in parallel, as to force players (both from the technical community and society) to keep in mind (and strive at) an integrative "solution", satisfying technical (passive safety) and societal needs at the same time. This combination sets it apart from other approaches, either purely technocratic along the conventional decide-announce-defend line or the voluntaristic policy some national programmes have reverted to in the face of the failed technocratic approach (e.g., Sweden, Japan, UK, USA) or any unsystematic negotiated mixed versions (Fig. 5.5). This concept of robustness is, further down, revisited and enriched by the notions of resilience and adaptiveness (Sects. 5.1 and 5.3). Based on international experience (Flüeler 2002d/2005c/2006b/2014b, d, 2015, 2016, 2021), a three-step approach for a site selection procedure is proposed, accompanied by an integrative assessment framework (Chap. 2). In Chap. 5, pertinent examples are given.

Step 1: Discuss—Comprehensive Societal Discourse
First, discuss the issue from all conceivable angles. "Involvement of stakeholders" cannot mean to call for as many individuals but for as many perspectives as possible to systematically incorporate all relevant facets in the

multi-dimensional discourse. The aim is to lay all pertinent aspects (values, norms, context, evidence …) open, to successfully "close" issues and proceed to the following phase. A "socially robust" procedure must, per se, allow for some flexibility and adaptability to an evolving social context, particularly given the long timeframe involved (decades to find a suitable site, to license, build, operate and duly close it). "Closing" steps, however, are crucial in order to avoid starting all over again. For example, fundamental issues like final disposal vs. retrievability of waste (and reversibility of decisions) have to be put on the table. This is sensitive, somewhat paternalistic (towards future generations), but it has to be done, in a comprehensive way.

> Most people do not listen with the intent to understand;
> they listen with the intent to reply.
>
> Stephen Covey, 2004

This long-time process has to be overseen, e.g., by a widely credible and trustworthy body. In 2002, Flüeler suggested a "National Council for the Safe [and Secure] Governance of Radioactive Waste" as the guardian of the process; the Swiss expert committee EKRA foresaw a "Disposal Council" (EKRA 2002). It should be pluralistically composed, independent of the industry yet knowledgeable and not driven by daily politics (Flüeler 2002d; Sects. 6.2.3, 7.1, 7.4, Fig. 7.1).

Step 2: Decide—"Common Ground" in Goals and Stepwise Strategy
The goals have to be prioritised so as to adopt the stakeholders' respective responsibilities. Conflicting goals exist (see, e.g., ENTRIA 2014). The concept of "sustainability", a complex goal indeed, encompasses protection of, and leeway for, future generations. Transferred to nuclear waste, both passive safety and "active" control or surveillance need due care and attention in parallel. No consensus will be reached "at heart", in the stakeholders' core beliefs. Society must agree on three levels as stated above.

Step 3: Implement—Start Programme and Prepare Long-Term Knowledge Basis
In view of a successful transfer of knowledge, it is vital to explore contextual issues and tacit/implicit knowledge— they determine the degree of societal understanding of the eventual disposition system. Unless the rationale of conceptual reasoning is appropriately handed over to the next—

technical, political and societal—generations, the entire undertaking is bound to fail (Flüeler 2005c).

The rules and criteria of site selection procedures have to be consented to before the start (in Step 1) and adhered to during the process. Revisions should undergo careful reviews and be consented to. A clear distinction between the implementer and regulatory bodies is vital. The regulators must establish a platform for inclusive knowledge generation, based on a (pre)defined set of criteria. This necessity to integrate different requirements, the step-by-step approach, the chance of "institutional constancy" and the special "national" task of the issue call *special attention to the role of the authorities* (NEA 2003). Issues like regulatory capture, expert blocking or technological lock-ins have to be duly considered (Chaps. 5 and 6).

Pour voir clairement, il suffit souvent de changer la direction de la vue. *In order to see through it is often enough to change the line of vision.*

Antoine de Saint-Exupéry (1900–1944), Citadelle, 1948

3.5   Conclusions and Summary

The three waste disposal systems—nuclear repositories, long-term landfills for other toxic waste and CCS—indisputably are controversial sociotechnical issues. Based on an amplified notion of sustainable development, fundamental rules are suggested to follow, in order to eventually find "common ground" (with a common goal) to reach respective long-lasting "solutions". This necessitates a comprehensive societal dialogue to identify major open issues and possible ways to go in parallel. Society must agree on three levels: first, on the problem recognition (waste exists, problem must be tackled, at least set on track to be solved); second, consensus on goals (degree of protection and intervention must be defined, with passive safety as top priority); lastly, consensus on the procedural strategy (with "rules of the game" clear from the outset).

Questions

1. What is the benefit of the concept of multiperspectiveness?
2. How would you distinguish "strong" from "weak" sustainability?

3. Why would a proponent of a technical facility want to engage in a "societal dialogue"?

Answers

1. To put oneself in someone else's position helps understand his or her arguments and maybe even their values behind (Fisher et al. 1981/1991). It furthermore is a tool to reflect our own arguments and views. Among other ramifications (beyond the scope of this contribution), it may also surpass the conventional linear communication model of "sender —message—receiver" (whose exclusive success criterion is whether the receiver understands or not the sender's message). In reality, when communicating, we are in a biological, psychological and cultural context which (co-)shapes all actors and messages.
2. "Strong" sustainability argues that some components and functions of the environment must not be substituted, e.g., the ozone layer cannot be replaced by anything (ecology is dominant). "Weak" sustainability, however, postulates that "human" capital may substitute "natural" capital, e.g., profits from clearing tropical forests may improve infrastructure (schools, health system) which is handed over to the next generations.
3. The three policy fields under scrutiny have, overall, not really been success stories so far. Neither the "technical" approach nor the "acceptance" approach are sufficient (Goldblatt et al. 2012). In the long run, it is wise to know where the problems and the expectations lie and to learn from them. This requires the integration of various, the relevant, perspectives in society.

Exercise 3.1
Figure 3.2 is dedicated to the radioactive waste system. Create a similar figure for conventional waste and CCS. What is common? What are the differences?

Exercise 3.2
Apply the three-step approach to different national radioactive waste programmes in groups (e.g., Sweden, Finland; Canada; France; Germany, Switzerland; United Kingdom; USA).

Exercise 3.3

Apply and adjust the three-step approach to the conventional waste and CCS systems. Carve out similarities and differences (e.g., retrieval for CCS is rather impossible).

Additional Information

All weblinks accessed 27 January 2023.

Key Readings

Key References

Flüeler T (2001a) Options in radioactive waste management revisited: a framework for robust decision making. Risk Anal 21(4):787–799. https://doi.org/10.1111/0272-4332.214150

Flüeler T (2004a) Long-term radioactive waste management: challenges and approaches to regulatory decision making. In: Spitzer C, Schmocker U, Dang VN (eds) Probabilistic safety assessment and management 2004. PSAM 7–ESREL '04. Berlin, June 14–18, vol 5. Springer, London, pp 2591–2596. https://link.springer.com/chapter/10.1007%2F978-0-85729-410-4_415

Flüeler T (2004b) etc.: other own references in Annex

Robustness, Production of Knowledge

Jasanoff S (1995) Beyond epistemology: relativism and engagement in the politics of science. Soc Stud Sci 26:393–418. https://www.jstor.org/stable/285424

Rammert W (2016) Technik – Handeln – Wissen. Zu einer pragmatistischen Technik- und Sozialtheorie. Springer, Wiesbaden

Rip A (1987) Controversies as informal technology assessment. Knowl: Creation, Diffus, Utilization 8(2): 349–371. https://doi.org/10.1177/107554708600800216

Schumpeter JA (1942, 1994) Capitalism, socialism and democracy. Routledge, London. https://www.routledge.com/Capitalism-Socialism-and-Democracy/Schumpeter/p/book/9780415107624

Sombart W (1916ff, 2012) Der moderne Kapitalismus. Historisch-systematische Darstellung des gesamteuropäischen Wirtschaftslebens von seinen Anfängen bis zur Gegenwart. Historisches Wirtschaftsarchiv. Orell Füssli, Zürich

Consensus, Conflicts, Participation, Etc.

Berkhout F (1991) Radioactive waste. Politics and technology. Routledge, London, p 12. https://www.routledge.com/Radioactive-Waste-Politics-and-Technology/Berkhout/p/book/9780415054935

Carter LJ (1987) Nuclear imperatives and public trust: dealing with radioactive waste. Resources of the Future, Washington, DC, p 427. https://doi.org/10.1177/0270467688008004156

ENTRIA (2014) Memorandum zur Entsorgung hochradioaktiver Reststoffe [Memorandum on the disposal of high-level radioactive residues]. In: Röhlig KJ et al Niedersächsische Technische Hochschule, Hannover, p 43. https://www.itas.kit.edu/english/projects_hock13_entria.php

Fisher R, Ury W, Patton BM (1981, 1991, 2011) Getting to yes. Negotiating an agreement without giving in. Houghton Mifflin, Boston MA/Penguin Books, New York

Goldblatt DL, Minsch J, Flüeler T, Spreng D (2012) Introduction (Chap. 1). In: Spreng D, Flüeler T, Goldblatt DL, Minsch J (eds) Tackling long-term global energy problems: the contribution of social science. Environment & Policy, vol 52. Springer, Dordrecht NL, pp 3–10. https://link.springer.com/chapter/10.1007/978-94-007-2333-7_1

Herzig EB, Statham ER (1993) When rationality and good science are not enough: science, politics and the policy process. In: Herzig EB, Mushkatel AH (eds) Problems and prospects for nuclear waste disposal policy. Greenwood, Westport, CT, p 10. https://www.goldenlabbookshop.com/book/9780313290589

OECD (2020) Innovative citizen participation and new democratic institutions: catching the deliberative wave. OECD Publishing, Paris. https://doi.org/10.1787/339306da-en

Sabatier P (1987) Knowledge, policy-oriented learning, and policy change. Knowl: Creation, Diffus, Utilization 8(4):649–692 https://doi.org/10.1177/0164025987008004005

Mental Models, Paradigms

Jones NA, Ross H, Lynam T, Perez P, Leitch A (2011) Mental models: an interdisciplinary synthesis of theory and methods. Ecology Soc 16(1):46. http://www.ecologyandsociety.org/vol16/iss1/art46/

Klose W (1994) Ökonomische Analyse von Entscheidungsanomalien. European university studies. Series V. Economics and management, vol 1533. Peter Lang, Frankfurt aM/New York

Kuhn TS (1962, 1970, 2012) The structure of scientific revolutions. University of Chicago Press, Chicago. https://press.uchicago.edu/ucp/books/book/chicago/S/bo13179781.html

Rohrmann B (1991) Akteure der Risiko-Kommunikation. In: Jungermann H, Rohrmann B, Wiedemann PM (eds) Risikokontroversen. Konzepte, Konflikte, Kommunikation. Springer, Heidelberg, pp 355–370

Schein EH (1985, 2016) Organizational culture and leadership. Jossey-Bass, San Francisco. https://www.wiley.com/en-us/Organizational+Culture+and+Leadership%2C+5th+Edition-p-9781119212058

Wiedemann PM (1991) Strategien der Risiko-Kommunikation und ihre Probleme. In: Jungermann H, Rohrmann B, Wiedemann PM (eds) Risikokontroversen. Konzepte, Konflikte, Kommunikation. Springer, Heidelberg, pp 371–394

Sustainable Development

Grunwald A (2016) Nachhaltigkeit verstehen. Arbeiten an der Bedeutung nachhaltiger Entwicklung. oekom, München (Box 4.1)

Howarth RB (1992) Intergenerational justice and the chain of obligation. Environ Values I:133–140

Majone G (1989) Evidence, argument, and persuasion in the policy process. Yale University Press, New Haven, p 169

Socolow RH, English MR (2011) Living ethically in a greenhouse. In: Arnold DG (ed) The ethics of global climate change. Cambridge University Press, Cambridge, pp 170–191

Solow RM (1974) The economics of resources and the resources of economics. Rev Econ Stud 41:29–45

WCED, World Commission on Environment and Development (1987) Our common future. Brundtland report. Oxford University Press, Oxford, p 46. https://sustainabledevelopment.un.org/content/documents/5987our-common-future.pdf

Nuclear Waste

EKRA (2002) Beitrag zur Entsorgungsstrategie für die radioaktiven Abfälle in der Schweiz. Im Auftrag des Departements für Umwelt, Verkehr, Energie und Kommunikation (UVEK) [Contribution to the Swiss disposal strategy for radioactive waste on behalf of the Ministry of Environment]. Bundesamt für Energie, Bern

IAEA, International Atomic Energy Agency (2000) Measures to strengthen international cooperation in nuclear, radiation and waste safety. International Conference on the Safety of Radioactive Waste Management. Córdoba, 31 May 2000. IAEA, Vienna. https://www.iaea.org/publications

Kojo M, Richardson, PJ (2009) The role of compensation in nuclear waste facility siting. A literature review and real life examples. Deliverable D16b. Argona project. https://cordis.europa.eu/project/id/36413/de

NEA, Nuclear Energy Agency (1995) The management of long-lived radioactive waste. The environmental and ethical basis of geological disposal of long-lived radioactive wastes. A collective opinion of the radioactive waste management committee. OECD, Paris. https://www.oecd-nea.org/jcms/c_12892/radioactive-waste-management

NEA (1999) Confidence in the long-term safety of deep geological repositories. Its development and communication. OECD, Paris

NEA (2003) The regulator's evolving role and image in radioactive waste management. Lessons learnt within the NEA Forum on Stakeholder Confidence. OECD, Paris

NEA (2012) Actual implementation of a spent nuclear fuel repository in Sweden: seizing opportunities. Synthesis of the FSC national workshop and community visit, Östhammar, Sweden, 4–6 May 2011. NEA/RWM/R(2012)2. OECD, Paris

NEA (2013) The nature and purpose of the post-closure safety cases for geological repositories. NEA/RWM/R(2013)1. OECD, Paris

North DW (1999) A perspective on nuclear waste. Risk Anal 19(4): 751–758, 751

Parker FL, Broshears RE, Pasztor J (1984) The disposal of high-level radioactive waste 1984. A comparative analysis of the state-of-the-art in selected countries. Swedish National Board for spent nuclear fuel. Beijer Institute, Royal Swedish Academy of Sciences, Stockholm, vol I + II. NAK Rapport 11:101

Swiss Government, Schweizerischer Bundesrat [Federal Council] (1957) Botschaft über die Ergänzung der Bundesverfassung durch einen Artikel betreffend Atomenergie und Strahlenschutz. Bundesblatt 19:1137–1158

Websites

https://sdgs.un.org/topics (UN platform on sustainable-development issues)

https://en.wikipedia.org/wiki/Sustainable_development#Education_for_sustainable_development (introduction to sustainability/sustainable development)

https://www.oecd.org/gov/innovative-citizen-participation-and-new-democratic-institutions-339306da-en.htm (participatory approach)

Sources

Covey S (2004) The 7 habits of highly effective people. Powerful lessons in personal change. Free Press, Florence, MA (quote, Sect. 3.2)

Saint-Exupéry A (1948) Citadelle [The wisdom of the sands]. Gallimard, Paris (quote, Sect. 3.2)

Risk Characteristics and Evolution of (Risk and Safety) Concepts

4

Abstract

Basics of any policy analysis are the description of the characteristics of the respective waste systems, fulfilling the requirement of criterion #3, "Total-System Analysis and Safety Concept" (Sect. 2.3). The interplay of aspects of nuclear and conventional toxic ("special") waste as well as carbon dioxide storage, CCS, is investigated, using a novel integrated system assessment: material and system characteristics, risk assessment and regulatory approaches. The goal is to create profiles of strengths and weaknesses of waste systems that are similar in their risk characteristics but dealt with differently in risk management and regulation. A further objective is to draw lessons from the comparison of different discourses and procedures of waste with a similar profile with regard to decision-making processes (the reasons for different regulations of all three waste systems are not investigated here).

Keywords

Risk, risk mechanism • Robustness • Safety assessment • Long-term safety • Long-term governance • Radioactive/nuclear waste • Conventional toxic/hazardous waste • Carbon dioxide, CO_2 • Carbon Capture, Utilisation and Storage, CCUS • Disposal

Learning Objectives

- Understand the presented system assessment approach, with material and system characteristics, risk assessment and regulatory approaches;

- Know the differences in safety concepts of the three policy fields;
- Understand the concept of "robustness".

4.1 Approach

The features of nuclear and conventional toxic ("special") waste[1] are investigated, with a side glance to carbon dioxide storage, using a novel integrated system assessment elaborated in Flüeler 2006b/2012b/2013: material and system characteristics, risk assessment (this chapter) and regulatory approaches (Chap. 5). The goal is to create profiles of the strengths and the weaknesses of wastes that are similar in their risk characteristics but dealt with differently in risk management and regulation and, finally, governance (Chap. 6).

4.2 Waste … Waste?

As to the waste notion, there is no "scientific" or even object-based definition of what "waste" actually is, neither in the nuclear nor the special waste field, leaving CO_2 as being particular. What represents "waste" depends on its user or owner and is, therefore, basically a social construct—a painful statement for engineers, technicians and practitioners (Box 4.1). The boundaries between resources, products and wastes are blurred (Fig. 4.1). Thus, decision strategies and regulations cannot be separated from the (objective) respective waste system. This dynamic is hard to get used to but may also open up formerly unexpected paths towards "solutions".

---

[1] For key terms and concepts, refer to Glossary in the back.

© The Author(s), under exclusive license to Springer Nature Switzerland AG 2023
T. Flüeler, *Governance of Radioactive Waste, Special Waste and Carbon Storage*, Springer Textbooks in Earth Sciences, Geography and Environment, https://doi.org/10.1007/978-3-031-03902-7_4

a Conventional domain

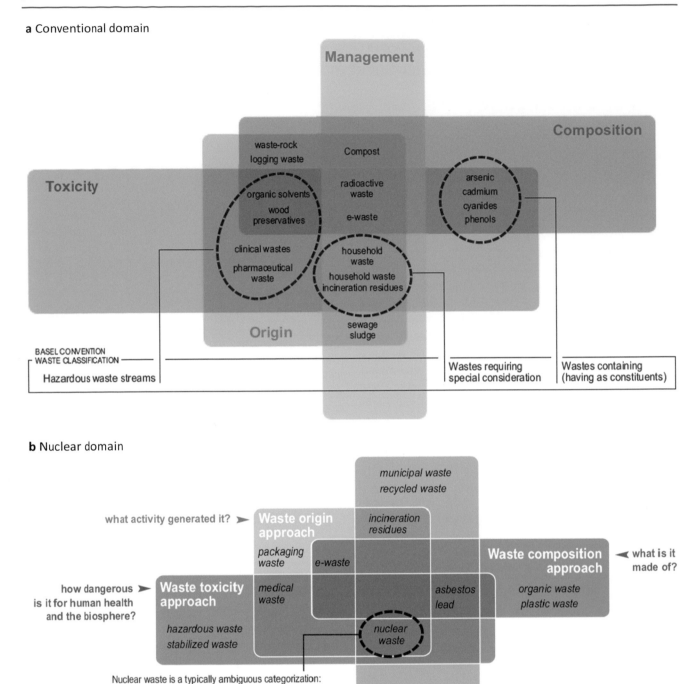

b Nuclear domain

Fig. 4.1 Waste substances of both domains are in a continuum even in their "technical" definition depending on the focus selected: composition, hazard potential, origin (polluters) or handling (reproduced from Flüeler 2013/2014, from UNEP 2004)

Box 4.1: Waste

Waste is defined by its owner. Waste is a resource in the "wrong" place or a residual to be disposed of (cf. Thompson 1979/²2017). Residual substances are toxic in the long term and have to be disposed of accordingly. See Fig. 4.1.

Legal text (ex.):
"Waste is any moveable material disposed of by its holder or the disposal of which is required in the public interest." (FEPA 2020, Art. 7, para 6)

"… in the end, these definitions [of waste] consist of lists of substances or groups of sustances and disguise the fact that there exists no *functional* definition which is not reduced to an enumeration but puts a waste *characteristic* at the forefront of the definition …. To attribute a waste characteristic to a substance or an object is the expression of a lacking esteem." (Grunwald 2016, 312–314, transl. tf, emphases original)

"Waste does not arise from physical, chemical or biological processes but from societal attribution of value – rather of: devalue." (Grunwald 2016, 323)

Nuclear/Radioactive Waste

Hazardous waste that contains radioactive material: with radionuclides, unstable isotopes of elements that decay and, by that, emit ionising radiation (material above clearance levels as established by the regulatory body and for which no use is foreseen). Nuclear waste may be divided into three classes: high-level waste (generating much heat by the decay process, from chemical reprocessing or spent fuel if it is declared a waste), intermediate-level waste, low-level waste (not requiring shielding during handling and transportation, long-lived and short-lived with half-lives shorter than 30 years) (IAEA 1994b).

Waste is "material … for which no use is foreseen", a loose definition adopted in the nuclear community, and only for "legal and regulatory purposes" (IAEA 1995, 20).

Legal text (ex.):
"radioactive substances or radioactively contaminated materials for which there is no further use." (FRPA 2017, Art. 25, para 1)

Conventional Toxic Waste

Hazardous waste that does not contain radioactive substances. Special (toxic) wastes are "wastes whose environmentally friendly disposal requires specific technical and organisational measures due to their composition and characteristics" (FOEN 2003,

transl. tf). Toxic waste can contain chemicals, heavy metals, etc., and may be reactive, ignitable and corrosive. In the United States, these wastes are regulated under the Resource Conservation and Recovery Act (US RCRA 1976).

Legal text (ex.):
"Special waste means waste designated as special waste in the list of wastes …." (FADWO 2020, Art. 3, lit. c)

Carbon dioxide (CO_2)

Carbon dioxide is a greenhouse gas: a gas that absorbs and radiates heat. Increases in greenhouse gases have tipped the Earth's energy budget out of balance, trapping additional heat and raising Earth's average temperature. Carbon dioxide absorbs less heat per molecule than the greenhouse gases methane or nitrous oxide, but it is more abundant and it stays in the atmosphere much longer. It absorbs wavelengths of thermal energy that water vapour does not, which means it adds to the greenhouse effect in a unique way. Increases in atmospheric carbon dioxide are responsible for about two-thirds of the total energy imbalance that is causing Earth's temperature to rise (from NOAA 2020, https://www.climate.gov).

This "non-definition" of waste might be shrugged off as the waste community's business, not society's. We throw things away and reject a "return to sender". We simply do not care, and "the waste issue originates from a specific act of valuation at determined locations of … material flows" (Grunwald 2016, 314). This is why the Swiss legislator stepped in by stating that waste may be disposed of if this is "required in the public interest" (FEPA 2020, Box 4.1). The state does not call us "responsible" citizens to account but takes care of the waste itself (or by way of mandated parties).

4.3 Long-Term Issues

The main mechanism of all waste disposition systems under scrutiny is a low-level but long-term, chronic release into the environment; it may be described as a slow degradation of an open system with concurrent large uncertainties (Fig. 4.2). Such potential impacts on deep geological repositories are hard to detect with respect to location and time (except for some scenarios of human intrusion). These system characteristics make radioactive and conventional highly toxic waste disposal—and CO_2 storage—unique compared to other technical risks.

These features lead to the admission that the required long-term safety (for radioactive waste) "is not intended to

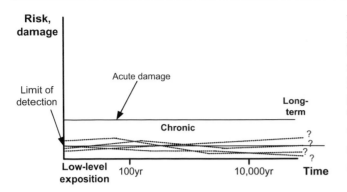

Fig. 4.2 System properties of toxic waste disposal. The impact of the waste is potentially chronic (no peak and no acute damage), perceived to be "creeping" (badly traceable; partly below the limit of detection, low-level dose or none in the case of carbon dioxide) and long-termed. Thereby, the actual low risk appears to be "smeared" over a long time span. No scale is given due to the generic character of the sketch. Dashed curves are results of possible release scenarios (reproduced from Flüeler 2006b, 17)

imply a rigorous proof of safety, in a mathematical sense, but rather a convincing set of arguments that support a case for safety" (NEA 1999, 11; NEA et al. 1991, 10ff). Due to the required longevity of the disposal system, "[t]he aim of the performance assessment is not to predict the behaviour of the system in the long term, but rather to test the robustness of the concept as regards safety criteria" (IAEA 1999, 245–246). Following this, sophisticated disposal concepts and designs were developed, with succeeding technical barriers, review cycles and quality assurance (e.g., NEA 1999). Building upon the defence-in-depth principle, the concept of integral, technical and societal, robustness was developed (Flüeler 2002d/2006a, b) and is reproduced and enlarged by "resilience" below.

4.4 System Characteristics

Regarding the waste notion, there is no "scientific" or even object-based definition what "waste" actually is, in neither the nuclear nor the special waste field. What represents "waste" depends on its user, processor and/or regulator, and may, therefore, vary in different countries and sociocultural contexts. The boundaries between resources, products and waste are fuzzy or—remember the problem discourse above —ill-defined. Thus, decision making and regulatory framing cannot be separated from the (objective) respective waste system (Tables 4.1 and 4.4).

4.4.1 Conventional Hazardous Waste/Radioactive Waste

In the conventional field, all substances have to be investigated with respect to their classification: as resource or

waste, e.g., batteries may have a positive market value if recycled, a negative if they are a legacy to be landfilled. Mercury can be reused as an extraction medium in gold mines or emitted into the environment. It has been valuable in clinical thermometers or batteries; it was, however, internationally banned by the Minamata Convention (2017). This is an example of substances that have to be phased out "forever", whereas others, like organic compounds, can be transformed in a way that is not harmful anymore (Baccini and Brunner 2012).

Substances (Table 4.2) burdened with radioactivity are cleared or decontaminated if possible. The remainder, nuclear waste is classified into around 160 types for conditioning and disposal (e.g., Nagra 2008). Conventional materials can be either completely recycled, burnt, conditioned or landfilled (in which latter case they are indeed considered waste). Special waste in particular can be divided into around 450 waste types, of which—in Switzerland— one-fifth is disposed of in a dedicated landfill (FOEN 2012, Fig. 5.8 for Europe). For the sake of completeness, mixed waste should also be mentioned here, meaning hazardous waste with radioactivity.

Regarding the **management principles**, the application of the sustainability concept suggests to close material cycles in both fields (Table 4.3). The ultimate closing of the fuel cycle has been the strategy in the nuclear industry ever since (NEA 2000) (even though current reprocessing merely extracting plutonium, partly uranium, for reuse, cannot be considered full recycling). Once waste is defined, it has to be kept away from humanity and the environment by disposal in landfills or repositories. In both fields, the principle of *confine and concentrate* has finally prevailed over *dilute and disperse* even though not always in reality and exceptions exist (e.g., the trade-off in ^{14}C treatment between real doses today and potential doses in the future; or natural microbial decomposition versus a rigorous concentration strategy). Both waste communities agree that waste is either a resource in the wrong place (not yet recycled) or residual material to be disposed of. Residuals are chronic and toxic in the long term and need safe long-term disposal. A circular economy strives to reuse and recycle substances (for maximum resource efficiency), whereas merely (according to the state of the art) persistent, non-recyclable material shall be deposited in final sinks, i.e., deposits (extended in Haberl et al. 2019; Kral et al. 2014).

Long-term **deposits** are usually above ground in the conventional domain, often—always in case of high-level waste—in deep geological underground in the nuclear domain. Both fields exhibit areas of transition, such as controllability and retrievability, on the one hand and no need for maintenance and ultimacy (in the sense of no intention for retrieval) on the other hand. Therefore, target conflicts are quite possible, as can be shown with the notion of retrievability, meant either in case of system failure or to

Table 4.1 Systemic and institutional properties of radioactive waste and conventional waste disposal. See the text for explanations (Seidl et al. 2021, Flüeler 2013/2014a)

Area	Radioactive waste	Conventional waste
Origin	Nuclear power plants (few), medicine, industry and research (many producers)	Large and heterogeneous production (households to industrial facilities)
Sources	Relatively few (Fig. 4.1b)	Many hazardous substances
Economy	Financing according to polluter-pays principle, PPP (per kWh)	Partly according to PPP (municipal waste bag fees, prepaid fee, suppliers' fee)
Resources	Many (R&D, repositories)	Partly (handling)
Actors	Relatively few producers, national regulatory agencies	Many producers (market companies, households), legislation and execution heterogeneous (partial states, international)
Volume (Switzerland) deposited	Small (conditioned) (100,000 t/60 y, high-level waste, HLW 3,000 t/50 y)	Large (often in bulk), special waste (100,000–200,000 t/y)
Concentration	Highly concentrated (high-level waste) Low (low-level waste)	Mostly diluted Locally concentrated
Waste package	Conditioned, packed	Dispersed, unpacked (fly ash), partly diluted, only locally concentrated
Nature of substance	Mostly solid waste (reprocessing: goods)	Mostly solid waste or goods
Regulatory classification	International, national	National, heterogeneous; with regard to characteristics and toxicity not suitable for deposits
Facilities	Dedicated mine (of various types), internationally partly abandoned mines	Landfills, deposits (of various types), exceptionally underground waste disposal
Range of options/flexibility	Relatively small (reprocessing, partitioning & transmutation?)	Large (substitution, reuse, recycling)

Notes Italics Commonalities of both systems. Underlined Advantages of conventional waste. **Bold** Disadvantages of conventional waste. PPP polluter pays principle, R&D Research and Development, t metric tonnes

have access to valuable material (Flüeler 2012a). Whereas in the radioactive field, there are basically only two (or three) repository types (for low-level and high-level waste), the conventional field undergoes a lively dynamic: partly by increasing deposition and partly by progress in recycling. Thus, there exists no accurate selectivity as to which *special waste* will be effectively deposited in the future or rather recovered. After decades of dumping, the notion of "final disposal" was accepted in the conventional waste field (Baccini 1989). Nowadays, the goal of advanced waste management is to only deposit non-recyclable substances, and, at that, in so-called safe last sinks with no or minimal emissions to the environment over the long term. Organics are burnt in incinerators as last sinks. "The need for final sinks can be reduced by means of substance bans, product lifetime extension and recycling. However, sinks cannot be prevented completely" (Kral et al. 2019, 8).

With respect to **quantity** (not toxicity), it is quite obvious who prevails. In Switzerland only, the annual accumulated special waste to be deposited in landfills corresponds to the total waste the five nuclear power plants produce during an operation period of 50 years (including expected nuclear waste from medicine, industry and research over a collection period of 80 years). In the conventional sector, however, the recycling rate is high and increasing, whereas the nuclear handling is limited to decontamination or possibly reprocessing (FOEN 2019; Nagra 2008).

A great number of **actors** and a multitude of regional regulatory bodies characterise the conventional domain (e.g., Wollmann 2015), whereas producers and users of nuclear material are small in numbers, so the competences are centralised and supervision rests with few, mainly federal, agencies. Even though there is a flourishing international waste business, the conventional community is usually regionally, at best nationally, connected regarding municipal solid waste and sewage sludge—the nuclear community, though, is closely internationally linked and institutionalised (with standardisation and review mechanisms).

Law, in general, bases **financing** both fields of application on the polluter pays principle (PPP, or principle of causality). It is evident, however, that this is easier to follow in the nuclear field with relatively few actors, waste types and a dedicated up-front producer-paid fund for disposal. The picture is much more complex in the case of conventional waste management. With municipal waste, the Swiss Federal Supreme Court ruled that all municipalities had to introduce a so-called bag fee; not more than 30% of the disposal costs are allowed to be paid by public funds, i.e., taxes. This PPP

Table 4.2 Risk characteristics of the systems "radioactive waste" and "conventional (hazardous) waste". See the text for explanations (Seidl et al. 2021, Flüeler 2013/2014a, slightly modified)

Area	Radioactive waste	Conventional waste
Hazardous substances	Relatively few, well characterised; well-known effect of mid-range and high-radiation doses, effect of low doses (within and below the range of natural background) topic of scientific debate	Many substances with diverse, partly badly known effects, especially when simultaneously present, natural or synthetic (e.g., pesticides, polychlorinated biphenyl, PCB, polycyclic aromatic hydrocarbons, PAC)
Effect	Basically identical effect of all radionuclides (α-, β- und γ-emitters, energy doses), though divergent biological behaviour	Partly threshold dose, partly linear dose–effect relationship discovered; indication of concentrations instead of potentially effective doses
Conditioning	Concentrated, consolidated waste stream, mineralised (HLW); partly organic (operational waste, medicine/industry/research)	Dispersed, unpacked (fly ash); according to environmental legislation minimised organics
Toxicity	High (low with short-lived low-level waste)	High and low
Duration of toxicity (residence time in systems)	Very long (millions of years) to short (minutes, hours)	Very long (unlimited) to none (transformation, substitution)
(Long-term) Hazard potential	Complex (ionising radiation)	Varying (antagonistic, additive, synergistic)
Occurrence	Compared to nature relatively high artificial concentration (with HLW)	Compared to nature relatively low artificial concentration
Risk level	Low, chronic	Medium (surface landfills), chronic; floating boundary between biological demand and negative impact (trace elements, e.g., zinc)
Main protection goals (values)	Human being, radiation dose (Millisievert)	Man and environment (water), LD-50/none
Indicators	Defined for safety functions (such as isolation, confinement, decay), comparable with toxicity	None (no thresholds), badly comparable

Notes Italics Commonalities of both systems. Underlined Advantages of conventional waste. **Bold** Disadvantages of conventional waste. HLW high-level waste, LD lethal dose

conforming sentence results in about one-third less municipal waste; this fraction is collected separately and largely recycled (FOEN 2012, 27). Wastes from the building industry and trade are organised by the private sector itself. Sewage sludges are paid by way of a domestic and industrial wastewater fee, respectively. Either a prepaid disposal fee covers special waste or the waste specialist directly charges a fee to the waste owner. Long-term safety in the waste treatment on landfills, however, is not covered by the waste producers but by the state. Concerning legacy waste (on contaminated sites or brownlands) in Switzerland, a separate fund exists provisioned by the respective landowners or, in case they cannot get hold of, the state.

The relatively small volume paired with well-equipped funds results in the positive situation that the nuclear community with few waste owners can spend a lot of resources on focused R&D.

(Nuclear) repositories and (conventional) landfills are both associated with a similar **risk mechanism**: a long-term chronic release of pollutants into the environment (Table 4.2). The performance assessment for nuclear waste is portrayed in Section 4.5. The comparable complexity of the disposition of special waste is not reflected in the practised risk methodology. In this domain, we are also confronted with timeframes of hundreds to hundred thousands of years. Swiss standard leaching tests, even though stringent in an international comparison, permit no information on the long-term behaviour of a disposition system, let alone long-term safety (FADWO 2020). The definition of **protection goals** for landfills is vague and quite general, whereas the nuclear domain works with calculable annual doses, which may be compared with a defined regulatory limit as the protection goal. The measured variable here is radioactivity, which usually can be quantified accurately, whereas the conventional domain largely refers to concentrations of contaminants. The heterogeneous composition of waste increases the risk to miss interactions of known and even new substances.

Apart from unintentional human intrusion, the main **risk mechanism of repositories, landfills (and contaminated sites**, for that matter) is a low-level but long-term chronic release into the environment; as above, it may be described as a slow degradation of an open system (geological formations and geotechnical barriers/weather conditions, respectively) with concurrent large uncertainties. Substances with chronic effects are hard to detect. In the nuclear field, they are quantified by means of an uptake of radioactivity; in the conventional field, however, concentrations of toxins are given in the absence of impact analyses.

In view of the recognised complex system mechanism, there is consensus in the nuclear community that for the required **long-term safety** of repositories no rigorous proof is strived for but convincing arguments supporting the "safety case" (Sect. 4.5.1). The Joint Research Centre, commissioned by the European Commission in 2018 to evaluate nuclear energy, came to the conclusion: "Although nuclear energy has been recognised … as 'climate-neutral energy', the compliance with the 'do no significant harm' criteria of the nuclear energy life-cycle, and in particular the disposal of radioactive waste, requires further considerations …. No long-term operational experience is presently available as technologies and solutions are still in demonstration and testing phase moving towards the first stage of operational implementation" (JRC 2021, 7f).

In the conventional field, the understanding of the—technical—system and the data basis are often insufficient. Input data (such as chemical and physical waste characteristics), but even more so long-term mechanisms (like multiple-phase thermodynamics), are not well known. Flux and transport models in unsaturated conditions are premature and partly difficult to obtain. The rough understanding of individual processes, to date, does not permit reliable safety analyses for above-ground disposition systems (Herrmann and Röthemeyer 1998, 297) even though progress was made (Townsend et al. 2015; Wang et al. 2021).

The situation is complex because a large number of materials comprise hazardous substances, as the European Commission, EC acknowledges: "The complexity of waste streams, recovery processes and recovered materials means that end-of-waste criteria that are applicable to whole waste streams are not easy to establish" (EC 2018, 5). Moreover, many powerful agents lobby against stricter compliance with regulations (e.g., Kraft 2017; Vogel 2012).

In consequence, the intention of circular economy is to reuse, recycle and recover substances to the fullest and to deposit merely refractory waste material in so-called safe final sinks (Kral et al. 2014). Within its Circular Economy Package, the European Union plans to phase out landfilling for recyclable waste (including plastics, paper, metals, glass and biowaste) in non-hazardous waste landfills by 2025 corresponding to a maximum landfilling rate of 25% of the total waste (EU 2019).

4.4.2 Carbon Capture and Storage

Some observers have argued that CCS is by no means a novel technological concept as CO_2, among other substances, has long been injected into oil formations to recover additional oil (Benson 2005). What is radically different are the respective goals and scales of CCS and enhanced oil recovery. The new intention with CCS is to seclude CO_2 for the common good (a sustainable climate, as it is harmless to individuals, even at high concentrations), at an ample scale and for a very long time. This requires the implementation of an internationally effective system, tested long-term safety performance of facilities and the overall system (including an adequate control mechanism) and, above all, the recognition of longevity as a factor to be dealt with.

When CO_2 is emitted, much of it remains in the atmosphere for around 100 years. The carbon cycle of the biosphere takes a long time to neutralise and sequester anthropogenic CO_2. According to Archer (2005), the **lifetime of fossil fuel CO_2** is 30,000–35,000 years (Archer 2005) if averaged out over its entire distribution. The term "lifetime" refers to the period required to restore equilibrium following an increase in the gas's concentration in the atmosphere. If one includes the long tail, Archer suggests that this figure is "300 years, plus 25% that lasts forever" (ib., 6). This, along with the heat capacity of the oceans, means that climate warming is expected to continue beyond 2100, even if CO_2 emissions cease (Fig. 4.3).

Figure 4.4 depicts the well-known, long-term dimension of nuclear waste, with a drop in radioactivity for low-level waste after 1000 years.

Comparing radioactive wastes with CO_2 reveals both differences and similarities, both in **systemic and risk aspects** (Tables 4.4 and 4.5, respectively). It is less about the sheer size of the release into the environment and has more to do with the nature of it (Benson 2005). Disposal and storage systems are both associated with a similar **risk mechanism**: a low-level but long-term chronic release of pollutants into the environment along with a slow degradation/alteration of an open system (geological formations and geological formations/climate, respectively) with concurrent large uncertainties. With the exception of some scenarios of human intrusion, potential impacts are hard to detect with respect to location and time. The notions of what "long term" may mean from different perspectives are scrutinised in Section 5.2.

With respect to **long-term safety**, the JRC came to the conclusion that, like nuclear, "similarly, carbon capture and sequestration (CCS) technology is based on the long-term disposal of waste in geological facilities and it has been included in the taxonomy and received a positive assessment. The Taxonomy Expert Group therefore considers that the challenges of safe long-term disposal of CO_2 in geological facilities, which are similar to the challenges facing disposal of high-level radioactive waste, can be adequately managed. There is already an advanced regulatory framework in place in the communities for both carbon dioxide storage and radioactive waste management (see "Annex 1"). In terms of practical implementation, there is currently no

Table 4.3 Risk concepts and decision-making patterns of the systems "radioactive waste" and "conventional (hazardous) waste". See the text for explanations (modified after Seidl et al. 2021; Flüeler 2013/2014a)

Aspects	Radioactive waste	Conventional waste
Risk appraisal		
Concept	Concentrate and confine	*Do.,* dilute and disperse
Mechanisms	Sorption, diffusion (advection)	Mineralisation (by way of preceding incineration), dispersion
System understanding	Existent, developed	Extremely divergent
Site characterisation	Good, dedicated investigations (seismics, boreholes, etc.), systematic procedures	Medium to bad (limited risk analyses, often no systematic site selection procedures)
Barriers: Technical Natural	Pellets, canisters/containers, bentonite Sealed galleries/access structures Host rocks and other isolating (confining) units	None/modest (inerting and monitoring) Sealings Caprock (only if underground waste disposal)
Assessment methodology	Relatively well developed ("safety case"/ "demonstration" of safety)	Highly variable
Uncertainties (main)	Scenarios (due to long term, e.g., erosion), human (failure, intrusion)	Modelling (sophistication), human (failure, intrusion)
Professional actors	Geologists, risk analysts, civil engineers	Civil engineers, risk analysts (geologists)
Paradigm (Switzerland)[a]	Deep geological disposal	Residue landfill and landfill for reactor waste; circular economy, final sinks
Other relevant issues		
Conflicts in land use	Existent (may be avoided)	Existent (may be avoided)
Process of demonstration of safety	Stepwise, in stages, iteratively	Undefined
Surveillance	Extended final disposal, optional/not necessary	Necessary (FADWO 2020: 50 y)
Range of options/flexibility	Small	Large but heterogeneous as well (substitution, mineralisation)

Notes Italics Commonalities of both systems. <u>Underlined</u> Advantages of conventional waste. **Bold** Disadvantages of conventional waste
[a] Other strategies are used in parallel (above-ground deposits with very low-level waste and so-called decay deposits for decommissioning waste until clearance of radioactive material)

operational geological disposal for carbon dioxide or for radioactive waste" (JRC 2021, 9). The review of the direct comparison of nuclear, conventional waste and CCS in that document (ib., 336–342), however, did not reveal conclusive evidence for the statement of "advanced regulatory framework" in CCS. The lack of regulations, national or international, in the CCS arena is mentioned in Chap. 2.

4.5   Safety Concept

4.5.1   Conventional Hazardous Waste/Radioactive Waste

In the radioactive waste community, evidence, analysis and arguments for the safety of radioactive waste disposal sites must be gathered in a so-called "safety case", which involves a stepwise and iterative procedure to provide risk assessments and appraisals, as well as confidence statements from site selection to closure and post-operational monitoring of a

specific facility (NEA 2008b). As specified below, the safety case for radioactive waste disposal envisions the following overall safety functions: isolation of the waste from the human environment, long-term confinement within the disposal system and attenuation of releases from the environment (IAEA 2003b, 5; IAEA 2006, 19ff). The key idea of compliance is the multiple-barrier concept, according to which a number of technical and natural barriers take effect in case one single barrier breaches (Fig. 4.5).

Following lengthy expert and public debate in the radioactive waste area over a period of several decades, final disposal in geological formations has emerged as the comparatively best solution (NEA 2020). Due to a rejection of the *disperse and dilute* concept, sea dumping was prohibited by the London Dumping Convention and, later, by the OSPAR Convention (LDC, OSPAR). Space disposition was dismissed and, likewise, sub-seabed disposal was abandoned for safety reasons, prohibition by maritime law and lack of public acceptance (NEA 2008a, c; OSPAR 2007). Although this is far from a "closed issue", the scientific and

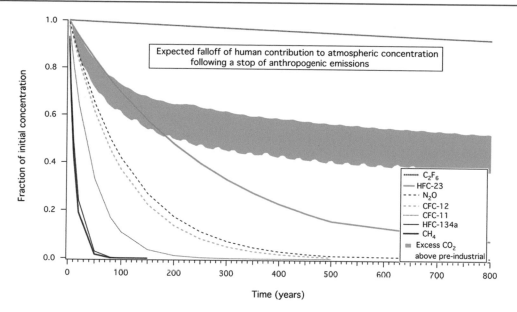

Fig. 4.3 Falloff of concentrations expected for various greenhouse gases, from a peak value following cessation of emissions. The decline of CO_2 to a high and stagnant level is notable (reproduced from Flüeler 2012d, from Solomon et al. 2009, supplementary material)

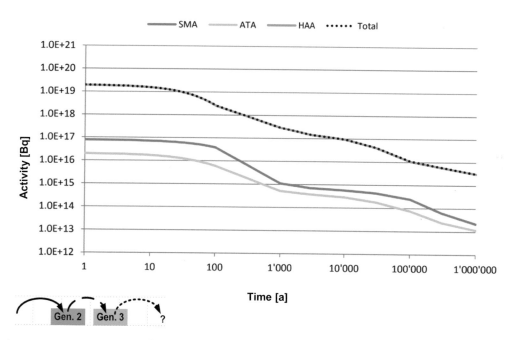

Fig. 4.4 Comparison of radioactive waste streams with respect to their radioactivity (measured in Becquerel, Bq). The reference is the Swiss nuclear programme with 50 years' history of operating five nuclear power plants with 3000 MW_{el} power and waste from medicine, industry and research accumulated over 70 years (until 2050). Due to this characteristic, geological barriers are chosen to isolate the waste from the biosphere. HAA: high-level waste and spent fuel (3,000 metric tonnes), SMA: low-level and short-lived intermediate-level waste, ATA: (intermediate) alpha-toxic waste ATW (reproduced from Nagra 2014, 34)

non-scientific debate is more advanced in the radioactive waste discourse than in toxic waste and even more in CCS (Tables 4.3 and 4.5, respectively). According to Bijker and colleagues, **"closure"** in science occurs "when a consensus emerges that the 'truth' has been winnowed from the various interpretations" (1987, 12). The differences in advancement to other fields are most striking in the safety assessment methodology and procedure, as shown below.

Safety analyses, which are usually called "performance assessments" in the context of radioactive waste, include scenario development, conceptual and mathematical models development, consequence analysis, uncertainty and

Table 4.4 Systemic and institutional properties of radioactive waste disposal and CO_2 storage. The large family of technologies is an <u>asset of CO_2</u>, whereas the absence of economical incentives and disposal funds, the high number of actors (polluters and potential implementers), the uncertain regulation and classification, the high volume and the non-dedicated type of facilities are **burdens to sustainable CO_2 storage**. See the text for explanations (Flüeler 2012b/2014)

Characteristic	System	
	Radioactive waste disposal	(Fossil) CO_2 CCS
Origin	Nuclear electricity generation; medicine, industry, research	Fossil-fuelled energy generation; oil refinery; cement, iron, steel industries
Sources	Comparatively few	Few large, innumerable small
Economy	Polluter-pays fee on each kWh	Not regulated to date
Resources (R&D, money)	Substantive	Sparse
Actors	Few state-by-state implementers; national regulatory agencies	Market-driven international companies; regulation undetermined
Volume	Small (Mt/60 yrs)	High (Gt/yr)
Waste package	Concentrated, consolidated waste stream	Dispersed
Nature of substance	Contaminant, waste	Greenhouse gas, commodity (waste)
Regulatory classification	Internationally classified	No standardised classification
Facilities	Dedicated repository with drifted galleries, abandoned mines substandard	Exploited reservoirs
<u>Range of options/flexibility</u>	Small	Large, from alternatives (e.g., renewables) to variants (e.g., ocean storage, mineralisation, etc.)

Notes Italics Commonalities of both systems. <u>Underlined</u> Advantages of CO_2 storage. **Bold** Disadvantages of CO_2 storage. Mt megatonnes, Gt gigatonnes

sensitivity analysis and confidence building. The central instrument is the development of a so-called **"safety case"**. According to NEA (1999, 22), "A safety case is a collection of arguments, at a given stage of repository development, in support of the long-term safety of the repository. A safety case comprises the findings of a safety assessment and a statement of confidence in these findings. It should acknowledge the existence of any unresolved issues and provide guidance for work to resolve these issues in future development stages". This implies that the point is not to start out by planning a repository in a manner that is as detailed as possible but to corroborate the safety case in steps that must be understood by all actors at every stage (NEA, id., 15).[2]

As for the barrier structure, the safety case distinguishes between the near field and the far field. The near field is defined as the engineered barrier systems, "as well as the region of rock immediately around waste emplacement tunnels (extending a few metres from the tunnels) [the Excavation Damaged Zone] that is significantly affected by thermal, hydraulic, chemical and mechanical changes induced by the presence of the waste and excavations. The far field is considered the [surrounding] host rock and the geosphere beyond [that is, other confining geological units], in which such effects [supposedly] are substantially smaller or negligible" (Nagra 2002, 111). System considerations include the evolution of climatic, surface environmental and geological boundary conditions, and they also involve "long-term processes such as uplift/erosion, climatic changes, neotectonic events, etc. The evolution of the near field and far field, driven by the presence of the waste and influenced by the boundary conditions, are then discussed [in Nagra 2002], including all aspects relevant to radionuclide transport" (Nagra 2002, 111). The consequence analysis is executed in a subsequent step, that is, within the site-specific safety analysis (for low-level radioactive waste, LLW, for example, Nagra 1994) and considers so-called "critical groups" that might be affected by the release of radioactivity within a defined scenario (Nagra 1994, 170).

The realisation, that is, the method of the safety analysis, consists of the following steps (as carried out by the Swiss implementer Nagra for the host rock Opalinus Clay and spent fuel, SF, high-level radioactive waste, HLW, as well as long-lived intermediate-level waste, ILW, according to Nagra 2002, IVf.):

---

[2] Different national programmes may apply the "safety case" approach differently and in a more detailed manner (SKB in Sweden, Andra in France) but generally quite similar (IAEA 2012).

Table 4.5 Risk-related properties of radioactive waste disposal and CO_2 storage. The longevity of pollutants, leakage as evasion mechanism, the low but chronic risk level and the long-term barrier mechanism are *common* characteristics. The relatively simple hazard system, good traceability and the wide range of options may be <u>assets of CO_2</u>, whereas most other waste and risk management aspects are at an initial phase and are therefore currently **burdens to sustainable CO_2 storage**. See the text for explanations (extended from Flüeler 2012b/2014a)

Characteristic	Radioactive waste	(Fossil) CO_2
P o l l u t a n t s		
Waste package	Concentrated, consolidated waste stream	None, dispersed
Toxicity	High (and low with short-lived low-level waste)	(Low but **dispersed**)
Duration of toxicity (residence time in systems)	Very long (millions of years)	Long (thousands of years)
<u>(Long-term) hazard system</u>	Complex (ionising radiation)	Simple (greenhouse gas effect)
Risk level	Low, chronic	Low, chronic
Protection limits	Radiation dose (Millisievert)	None (work safety: yes)
Indicators	Defined for safety functions	None (no thresholds)
<u>Traceability</u>	Bad (remote isolation)	Better
Transport mechanism	Groundwater, air (gas)	Air, groundwater
Evasion mechanism	Leakage	Leakage, seepage
R i s k a p p r a i s a l		
Concept	Concentrate and confine	(Confine)
Barrier mechanism	Sorption, diffusion	Mineralisation, dispersion
System understanding	Existent, developed	Under construction
Site characterisation	Good, dedicated investigations (seismics, boreholes, etc.)	Medium (lost records of abandoned reservoirs)
Barriers: **Engineered (technical)** **Natural**	Pellet, canister/container, bentonite; Sealed galleries and shafts Host rock and other confining geological units	None; Sealed wells Caprock(s)
Assessment methodology	Relatively well-established performance assessment ("safety case")	Partly advanced (seismics), partly embryonic
Uncertainties (main)	Scenarios (due to long term, e.g., erosion), human (failure, intrusion)	Modelling (sophistication), human (failure, intrusion)
O t h e r r e l e v a n t i s s u e s		
Use conflicts	Partly existent (can be avoided)	Often intrinsic (abandoned resource reservoirs)
Process of safety case	Stepwise, iterative	Undefined
In situ **monitoring**	Accessible shafts and galleries	Basically inaccessible (single-barrier seals)
Range of options/flexibility	Small	Large (variants, e.g., ocean buffering, mineralisation)

Notes Italics Commonalities of both systems. <u>Underlined</u> Advantages of CO_2 storage. **Bold** Disadvantages of CO_2 storage

1. The choice of a disposal system, via a flexible repository development strategy, that is guided by the results of earlier studies, including studies on long-term safety;
2. The derivation of the system concept, based on the current understanding of the features, events and processes, FEPs that characterise and may influence the disposal system and its evolution;
3. The derivation of the safety concept, based on well-understood and effective pillars of safety;
4. The illustration of the radiological consequences of the disposal system through the definition and analysis of a wide range of assessment cases;
5. The compilation of the arguments and analyses that constitute the safety case, as well as guidance for future stages of the repository programme.

The safety case, however, includes more information than just models and calculations of expected doses or risks. Additionally, it is based on the following broad lines of argument (Nagra 2002, excerpts from V, VIIIf. and Chap. 8):

- Strength of geological disposal as a waste management option. This is supported by (i) the internationally recognised fact that a well-chosen disposal system located at a well-chosen site fulfils the requirement of ensuring the safety and protection of people and the environment, as well as security from malicious intervention, now and in the future, (ii) the existence of suitable domestic rock formations, (iii) other safety assessments conducted worldwide, (iv) observations of natural systems and (v) the relative advantage of geological disposal over other options.

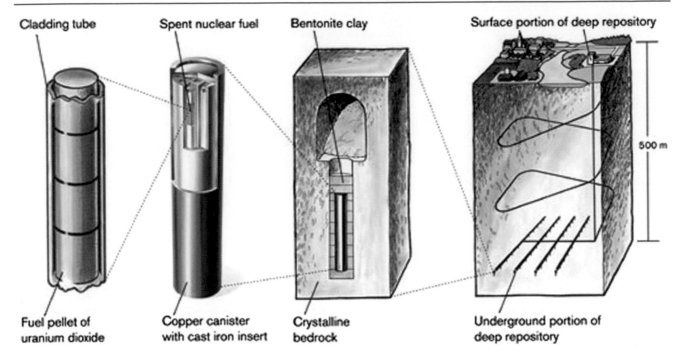

Cladding tube   **Spent nuclear fuel**   **Bentonite clay**   **Surface portion of deep repository**

500 m

Fuel pellet of   **Copper canister**   **Crystalline**   **Underground portion of**
uranium dioxide   **with cast iron insert**   **bedrock**   **deep repository**

Fig. 4.5 Multiple-barrier concept in radioactive waste management: the fuel pellets, the canister, the bentonite-clay buffer and a primary rock (such as the crystalline basement in the Swedish case) are designed to prevent the radioactive substances in spent fuel from spreading into the environment (reproduced from Flüeler 2012d, from SKB 2021)

- Safety and robustness of the chosen disposal system. This is ensured by (i) a set of passive barriers with multiple phenomena contributing to the safety functions, (ii) the avoidance of uncertainties and detrimental phenomena through an appropriate choice of site and design and (iii) the long-term stability of the chosen host rock and the repository due to a suitable geological situation.
- Reduced likelihood and consequences of human intrusion. This is supported by (i) the preservation of information about the repository, (ii) the avoidance of resource conflicts (that is, the absence of viable natural resources in the siting area) and (iii) the compartmentalisation of the repository and the solidification of the wastes.
- Strength of the stepwise repository implementation process. This is supported by (i) the fact that, at the current project stage, not all the details of the repository system need to be determined and, therefore, the information basis needs to be adequate for each particular stage, (ii) the reliance on understood and reliably characterised components (site, engineered barrier systems), (iii) the involvement of stakeholders and the opportunities for feedback and improvements, (iv) the flexibility of the project, allowing new findings to be taken into account (such as the detailed allocation of emplacement tunnels, choice of design options, placement of surface facilities and even siting; i.e., there are other possible sites for the chosen as well as for another host rock option) and (v) the

possibilities for monitoring and reversal (including waste retrieval).
- Good scientific understanding that is available and relevant to the chosen disposal system and its evolution. This is supported by (i) the results from regional and local field investigation programmes, key elements of which include an extensive 3D seismic campaign and multiple boreholes within each potential siting area and the availability of information from other investigations in the region, as well as complementary studies performed in an underground rock laboratory (Swiss: Mont Terri) and in other laboratories, as well as observations of the host rock in a number of railway and road tunnels, (ii) the findings from decades of experience in developing and characterising engineered barrier system components, as well as by the availability of a strong international information basis and (iii) the availability of a detailed model waste inventory for SF, HLW and ILW.
- Adequacy of the methodology and the models, codes and databases that are available to assess radiological consequences for a broad spectrum of cases. This is supported by adherence to the assessment principles outlined in Chap. 2 (of Nagra 2002).
- Multiple arguments for safety that include compliance with regulatory safety criteria, the use of complementary safety indicators, the existence of reserve FEPs and the lack of outstanding issues that have the potential to compromise safety.

4.5.2 Conventional Toxic Waste

In view of the recognised complex system mechanism, there is consensus in the nuclear community that for the required long-term safety of repositories it is recognised that a safety "proof" is not possible, instead one aims at a "convincing set of arguments that support a case for safety" (NEA 1999, 11; IAEA et al. 1991, 10ff). In the conventional field, the understanding of the—technical—system and the data basis are often insufficient. Input data (such as chemical-physical waste characteristics), but even more so long-term mechanisms (like multiple-phase thermodynamics), are not well known. Flow and transport models in unsaturated conditions are immature and partly difficult to obtain though progress has been made in rock labs (Bossart and Milnes 2018). The rough understanding of individual processes, to date, does not permit reliable safety analyses for above-ground disposition systems like shallow landfills (Herrmann and Röthemeyer 1998, 267ff). The safety design is depicted in Fig. 4.6.

A conventional "final repository" as defined in a seminal paper in the 1980s is a "deposit whose material fluxes into the environment (air, water and soil) are environmentally friendly without a follow-up treatment, and this in the short as well as long term" (FWC 1986, 37, transl. tf). The licensed and operating landfills in Switzerland do not comply with this principle; they merely fulfill the requirements of the valid waste ordinance FADWO: they are "waste facilities where wastes are deposited in a controlled way"[3] (Art. 3, lit. k).

4.5.3 Carbon Capture and Storage

The defence-in-depth paradigm central to nuclear installations is missing in the CO_2 safety concept for geological storage (Fig. 4.7). The basic idea here is that caprocks, such as aquitards, must prevent CO_2 from leaking back through sealed abandoned or dedicated injection wells or other pathways into the atmosphere. The dependence on what is basically one barrier is risky even if potential storage locations, such as—relatively few—depleted oil and gas reservoirs, are proven long-term traps of gas. Following a site selection and characterisation procedure according to the state of the art, the prerequisites for CO_2 sequestration are the reservoir (e.g., a saline aquifer, with sufficient porosity and permeability) and the caprock as a seal. Concerning caprock, a lot of research is still needed (are caprocks capable to form a tight seal for more than 10,000 years? What are the risks?). Is there fault integrity in relation to increased

seismicity when CO_2 penetrates tectonic fractures? Is there borehole integrity in old wells, CO_2 flowing back into the biosphere along leaking wells? What about a contamination of the biosphere, e.g., pollution of drinking water aquifers? A respective research example is CS-D experiment in the Mont Terri rock laboratory, where such risks are estimated by means of experiments (Zappone et al. 2021).

Waste and risk management for the CO_2 case both exhibit an early stage of research and no maturity of the safety concept (Table 4.5), even though progress has been made (Maul et al. 2007; Li and Liu 2016; https://www.globalccsinstitute.com) and shortcomings have been recognised (Bachu 2000). The wide range of options—from ocean storage to final trapping in minerals—is ambiguous for the appraisal of the "soundness" of the CO_2 safety conception. On one hand, it demonstrates the flexibility of waste management, while, on the other, it indicates a lack of maturity in the discourse of options.

The CCS community is far from an intricate buildup of scientific evidence and confidence as in the nuclear area, even though the issue has been recognised and a degree of international harmonisation is under way (IEA Greenhouse Gas R&D Programme) and science has become institutionalised (for example, by creating a dedicated journal, the International Journal of Greenhouse Gas Control, Gale et al. 2007). The needed technical systems go beyond the admittedly well-known injection of CO_2 but must still encompass the aspects described above. This means that CCS technology is not yet "mature", as maintained earlier (Bachu 2000, 957). Maul and colleagues recognised the following research demands in CCS: "Dealing with the various types of uncertainty, using systematic methodologies to ensure an auditable and transparent assessment process, developing whole system models and gaining confidence to model the long-term system evolution by considering information from natural systems. An important area of data shortage remains the potential impacts on humans and ecosystems" (Maul et al. 2007, 444). Grünwald (2008) and Jacobson (2009) list the research demands on leakage phenomena (nature, timing, mechanisms, etc.), while Savage et al. (2004) add scenario development to the agenda. The given status is still correct (Li and Liu 2016).

It is symptomatic that the term used is **"storage"** rather than **"disposal"**. In the nuclear community, the term "storage" always refers to temporary storage, whereas disposal is meant to be final in the sense that there is no intention of retrieval and that, if successful, (most) radionuclides will be kept out of the biosphere (IAEA 2006, 1). Retention is also possible by building CO_2 into minerals (mineralisation and carbonation), but this has only been implemented on a pilot-study level, not as state of the art. Natural analogues are also utilised to support a case (e.g., in the Triassic the CO_2 concentration was many times higher than it is today, and the dynamic processes can be derived from the rock archive of this period). Currently, there are only two commercial CO_2

---

[3] The official English translation (erroneously) reads: "landfill means a waste disposal facility in which waste is deposited", whereas the original German legal text says, "Deponien: Abfallanlagen, in denen Abfälle kontrolliert abgelagert werden" (FADWO 2020).

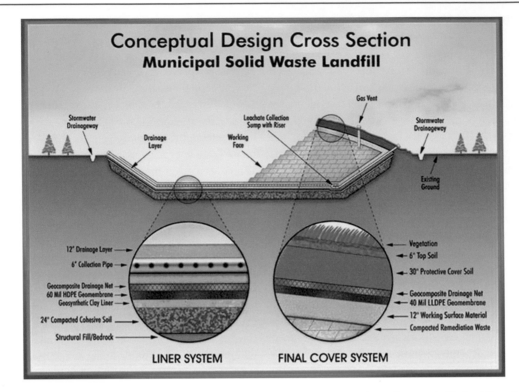

Fig. 4.6 Cross-section of a standard landfill with a liner system (geomembrane) to protect the underground ("bedrock") and a seepage water collection pipe (reproduced from Projectdataresearch 2022)

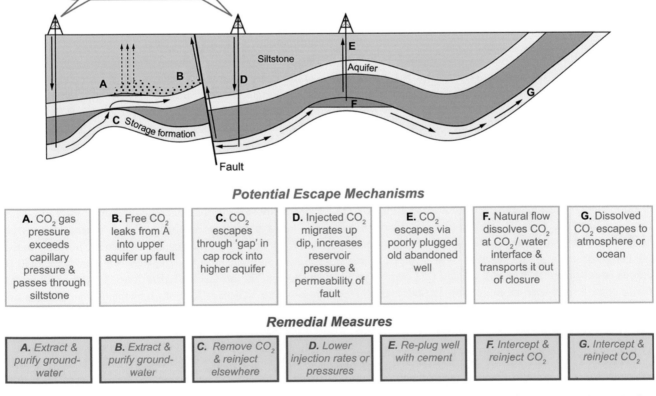

Fig. 4.7 Safety concept of CO_2 storage: CO_2 gas is injected into a "storage formation" as the only barrier (reproduced from Flüeler 2012d, from IPCC 2005, 35)

plants worldwide (retrofitted coal-fired plants), at Boundary Dam, Saskatchewan, and Petra Nova, Texas, USA, with a nominal capacity of 2.4 Mt/year (IEA 2020). The IEA concludes that this is "well off track to reach the 2030 [Sustainable Development Scenario] level of 310 Mt CO_2 per year" (ibid.). Benson (2005) expected CO_2 to spread at a rapid pace: 8.6 Mt injected annually would produce a plume of 18 km^2 over 30 years. Recent analyses show indeed intensified RD&D activities in the CCS community (Gomes and Corrazza 2019), but they concentrate on the incumbent oil companies' core business, the enhanced oil recovery, EOR. Even these powerful and competent actors do not push the CCS case forward. It is still plausible what Leung and colleagues concluded: "There are multiple hurdles to CCS deployment that need to be addressed in the coming years, including the absence of a clear business case for investment in CCS, and the absence of robust economic incentives to support the additional high capital and operating costs associated with CCS" (Leung et al. 2014, 439). Thus, it is only evident that states are called in for help (Romasheva and Ilinova 2019). Apart from "in house" arguments such as decarbonisation in hard-to-abate industries, the CCS field seems to concentrate on hydrogen production and on negative emissions with bioenergy (GCCSI 2020).

4.6 "Robustness"—The Technical Answer to (Supposedly) Cover All

There is wide international consensus that long-term radiation safety should not depend unduly on active measures. Hence, protection should be implemented primarily at the design stage (IAEA 1999, 243). Due to the required longevity of the disposal system, "[t]he aim of the performance assessment is not to predict the behaviour of the system in the long term, but rather to test the **robustness** of the concept as regards safety criteria" (ibid., 245). As a matter of course, robust procedures as defined in a narrow sense can only be achieved if the parameters are clearly defined and if it is assured that the system rests within well-set boundaries (Weinmann 1991, 33). Yet, the system characteristics of radioactive—and chemically toxic—waste are unique and technically complex. Once defined, the waste is to be stored in a safe way since it emits hazardous ionising radiation. Depending on the hazard potential of the waste in question its isolation period from the biosphere ranges from 100s to 100,000s of years.

As regards **technical robustness**, NEA makes the following distinction (NEA 1999, 30–37):

- *Engineered robustness:* "[i]ntentional design provisions that improve performance" such as overbuilding of barriers, waste conditioning in a stable matrix and physical separation of waste into packages of limited size;

- *Intrinsic robustness:* "[i]ntentional siting and design provisions that avoid detrimental phenomena and the sources of uncertainty" such as siting in sedimentary layers deep underground, with self-healing properties and an uneventful geological history, away from potential natural resources;

- *(Technical) system robustness:* combination of siting and design provisions supplemented by peer-review and quality assurance procedures.

A prediction of detriments to health over very long time periods is critical. Therefore, it is useful to consider safety indicators other than dose and risk criteria (IAEA 1994a, 2003a). This approach leads to:

- *Performance robustness:* comparison of the anthropocentric criteria individual human dose and risk with waste and environmental safety indicators, which are ruled by less uncertainty than the aggregated radiological protection goals.

The technical components of robustness, as envisaged by the international nuclear community, are depicted in Fig. 4.8. The amplification towards "integral" robustness and, beyond even that, to "resilience" is suggested in Chap. 5.

4.7 Conclusions and Summary

Science cannot define what waste is, in either field. Thus, decision strategies and regulations cannot be separated from the "objective" waste system. In the conventional domain, all substances have to be investigated in view of their classification (as resource or waste). Waste is a resource in the "wrong" place or a residual to be disposed of. Residual substances are toxic in the long term and have to be disposed of accordingly. Procedural and process issues make it explicit that perspectives (and their changes) and dynamics play an important role. A clear regulation of parameters and requirements as well as competencies and responsibilities of the involved parties are central. Independent regulators, on par with the implementer, must accompany and supervise all relevant steps. They need respective competences, resources and staff. It is wise to adequately involve third parties—and the larger public—at an early stage.

Comparing radioactive, conventional toxic wastes as well as CO_2 storage CCS reveals differences and similarities, both in systemic and risk aspects. Disposal and storage systems are both associated with a similar risk mechanism: It is expected that a low-level, but long-term, chronic release of pollutants into the environment will happen, along with a slow degradation/alteration of an open system (geological formations and geological formations/climate, respectively) with

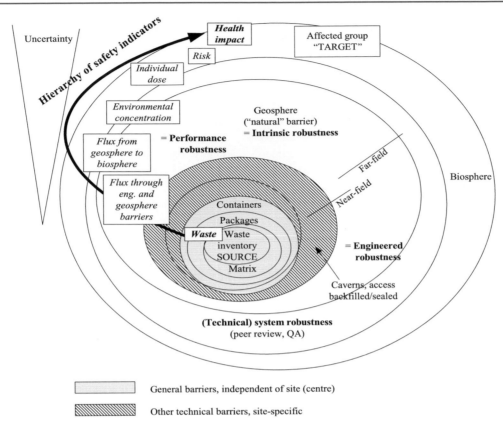

Fig. 4.8 Technical robustness. Long-term safety of long-lived radioactive waste is primarily based on passive technical (general, site-specific) or natural barriers. Technical robustness should be obtained through a careful selection of the site (intrinsic), prudent design of the disposal system (engineered), technical system robustness (reviewing, quality assurance, QA) and performance robustness. This last type is characterised by safety indicators other than dose and risk criteria (these are associated with higher uncertainties than waste or environmental measures) (reproduced from Flüeler 2006b, 235)

concurrent large uncertainties. With the exception of some scenarios of human intrusion, potential impacts on deep geological repositories are hard to detect with respect to location and time. In the radioactive waste community, evidence, analysis and arguments for the safety of radioactive waste disposal sites must be gathered into a so-called "safety case", which involves a stepwise and iterative procedure to provide risk assessments and appraisals, as well as confidence statements from site selection to closure and post-operational monitoring of a specific facility. The key idea of compliance is the **multiple-barrier concept**, according to which a multitude of technical and natural barriers take effect in case one single barrier breaches (though no complete redundancy can be achieved). This defence-in-depth paradigm, which is central to nuclear installations, is missing in the CO_2 safety concept for geological storage. The basic idea here is that caprocks, such as aquitards or any impermeable rock units, must prevent CO_2 from leaking back through sealed, abandoned or dedicated injection wells or other pathways into the atmosphere. The dependence on what is basically one barrier is risky even if potential storage locations, such as (relatively few) depleted oil and gas reservoirs, are proven long-term traps of gas.

Questions

1. What are the strengths and weaknesses of the three waste systems?
2. What is "waste"? Are ashes from incineration waste or not?
3. Carbon dioxide is odourless and nonpoisonous—what's all this fuss about?

Answers

1. Compare the Tables 4.1, 4.2, 4.3 and 4.4.
2. Waste is a resource in the "wrong" place or a residue to be disposed of. To date, filter, fly, or flue ashes from waste incinerators are dumped in a landfill. But with new technology, metals like zinc, lead and copper can be recovered as so-called "secondary resources"; they substitute "primary resources" which would otherwise have to be exploited as ores from mines. In parallel, noxes like mercury or cadmium or "forever chemicals" like per- and polyfluorinated alkyl substances, PFAS may be eliminated in order to reduce the hazard potential of landfills.

3. Carbon dioxide is even essential—as an "ingredient" of plant photosynthesis, without which humans could not live. But if it accumulates beyond a certain concentration in the atmosphere, it—somehow—also is a resource "in the wrong place".

Additional Information

All weblinks accessed 27 January 2023.

Key Readings

Key References

Flüeler T (2004a) etc.: other own references in Annex

Flüeler T (2012d) Technical fixes under surveillance—CCS and lessons learned from the governance of long-term radioactive waste management (Chap. 10). In: Spreng D, Flüeler T, Goldblatt D, Minsch J (eds) Tackling long-term global energy problems: the contribution of social science. Environment & Policy, vol 52. Springer, Dordrecht NL, pp 191–226. https://link.springer.com/chapter/10.1007/978-94-007-2333-7_10

Seidl R, Flüeler T, Krütli P (2021) Sharp discrepancies between nuclear and conventional toxic waste: technical analysis and public perception. J Hazard Mater 414:125422. https://doi.org/10.1016/j.jhazmat.2021.125422

Approach

Bijker WE, Hughes TP, Pinch T (1987) The social construction of technological systems. New directions in the sociology and history of technology. MIT Press, Cambridge, MA. https://mitpress.mit.edu/9780262517607/the-social-construction-of-technological-systems/

Weinmann A (1991) Uncertain models and robust control. Springer, Wien

Radioactive Waste and Conventional Toxic Waste

Baccini P (ed) (1989) The landfill. Reactor and final storage. Swiss Workshop on Land Disposal of Solid Wastes, Gerzensee, March 14–17, 1988. Conference proceedings. Lecture notes in earth sciences, vol 20. Springer, Berlin. https://doi.org/10.1007/BFb0011254

EC, European Commission (2018) Communication from the Commission to the European Parliament, the Council, the European Economic and Social Committee and the Committee of the Regions on the implementation of the Circular Economy Package: options to address the interface between chemical, product and waste legislation. Com(2018)32 final, Strasbourg

EU, European Union (2019) Waste framework directive. https://ec.europa.eu/environment/waste/target_review.htm

Haberl H, Wiedenhofer D, Pauliuk S, Krausmann F, Müller DB, Fischer-Kowalski M (2019) Contributions of sociometabolic research to sustainability science. Nat Sustain 2:173–184. https://doi.org/10.1038/s41893-019-0225-2

JRC, Joint Research Centre (2021) Technical assessment of nuclear energy with respect to the 'do no significant harm' criteria of Regulation (EU) 2020/852 ('Taxonomy Regulation'). RRC science for policy report. Sensitive. JRC124193. European Commission Joint Research Centre, Petten, NL

Kraft ME (2017) Environmental policy and politics. Routledge, New York. https://doi.org/10.4324/9781315437057

Kral U, Brunner PH, Chen PC, Chen SR (2014) Sinks as limited resources? A new indicator for evaluating anthropogenic material flows. Ecol Ind 46:596–609. https://doi.org/10.1016/j.ecolind.2014.06.027

Kral U, Morf LS, Vyzinkarova D, Brunner PH (2019) Cycles and sinks: two key elements of a circular economy. J Mater Cycles Waste Manage 21:1–9. https://doi.org/10.1007/s10163-018-0786-6

Townsend TG, Powell J, Jain P, Xu Q, Tolaymat T, Reinhart D (2015) Sustainable practices for landfill design and operation. Waste management principles and practice. Springer, New York. https://doi.org/10.1007/978-1-4939-2662-6_4

Vogel D (2012) Chemicals and hazardous substances. In: Vogel D (ed) The politics of precaution: regulating health, safety, and environmental risks in Europe and the United States. Princeton University Press, Princeton, pp 153–188

Wang LK, Wang MHS, Hung YT (eds) (2021) Solid waste engineering and management. Handbook of environmental engineering, vol 1 (23). Springer, Cham, Switzerland, p 3. https://doi.org/10.1007/978-3-030-84180-5_1

Carbon Capture and Storage, CCS

JRC, Joint Research Centre (2021) Technical assessment of nuclear energy with respect to the 'do no significant harm' criteria of Regulation (EU) 2020/852 ('Taxonomy Regulation'). RRC Science for Policy Report. Sensitive. JRC124193. European Commission Joint Research Centre, Petten, NL

Further Reading: Radioactive Waste

Bossart P, Milnes AG (eds) (2018) Mont Terri rock laboratory, 20 years. Two decades of research and experimentation on claystones for geological disposal of radioactive waste. Birkhäuser/Springer, Cham, Switzerland. See also https://doi.org/10.1007/s00015-016-0236-1

IAEA, International Atomic Energy Agency (1994a) Safety indicators in different time frames for the safety assessment of underground radioactive waste repositories. First report of the INWAC Subgroup on Principles and Criteria for Radioactive Waste Disposal. TECDOC-767. IAEA, Vienna. https://www.iaea.org/publications

IAEA (1994b) Classification of radioactive waste. A safety guide. Safety series 111-G-1.1. IAEA, Vienna

IAEA (1995) The principles of radioactive waste management. Safety fundamentals, Safety Series 111-F. IAEA, Vienna

IAEA (1999) Topical issues in nuclear, radiation and radioactive waste safety. Proceedings of an international conference. Vienna, 31 Aug–4 Sep 1998. IAEA, Vienna, pp 233–255

IAEA (2003a) Safety indicators for the safety assessment of radioactive waste disposal. Sixth report of the Working Group on Principles and Criteria for Radioactive Waste Disposal. TECDOC-1372. IAEA, Vienna

IAEA (2003b) Scientific and technical basis for the geological disposal of radioactive wastes. Technical Reports Series 413. IAEA, Vienna

IAEA (2006) Safety standards. Geological disposal of radioactive waste. Safety Requirements. No WS-R-4. IAEA, Vienna

IAEA (2012) The safety case and safety assessment for the disposal of radioactive waste for protecting people and the environment. Specific Safety Guide No. SSG-23. IAEA, Vienna

LDC, London Dumping Convention (1972) Convention on the prevention of marine pollution by dumping of wastes and other matter. https://www.imo.org/en/OurWork/Environment/Pages/London-Convention-Protocol.aspx

Maul PR, Metcalfe R, Pearce J, Savage D, West JM (2007) Performance assessments for the geological storage of carbon dioxide: learning from the radioactive waste disposal experience. Int J Greenhouse Gas Control 1(4):444–455. https://doi.org/10.1016/S1750-5836(07)00074-6

Nagra (1994) Bericht zur Langzeitsicherheit des Endlagers SMA am Standort Wellenberg (Gemeinde Wolfenschiessen, NW). Nagra Technical Report NTB 94–06. Nagra, Wettingen, Switzerland. https://nagra.ch/downloads

Nagra (2002) Project Opalinuston—safety report. Demonstration of disposal feasibility for spent fuel, vitrified high-level waste and long-lived intermediate-level waste (Entsorgungsnachweis). Nagra Technical Report NTB 02–05. Nagra, Wettingen, Switzerland

Nagra (2008) Mengen und Herkunft radioaktiver Abfälle [Quantities and origin of radioactive waste]. Nagra, Wettingen, Switzerland

NEA, Nuclear Energy Agency (1999) Confidence in the long-term safety of deep geological repositories. Its development and communication. OECD, Paris. https://www.oecd-nea.org/jcms/pl_13274/confidence-in-the-long-term-safety-of-deep-geological-repositories?details=true

NEA (2000) Nuclear energy in a sustainable development perspective. OECD, Paris

NEA (2008a) Moving forward with geological disposal of radioactive waste. OECD, Paris

NEA (2008b) Safety cases for deep geological disposal of radioactive waste: where do we stand? OECD, Paris. https://www.oecd-nea.org/jcms/pl_14342

NEA (2008c) Moving forward with geological disposal of radioactive waste: an NEA RWMC collective statement. NEA/RWM(2008)5/Rev2. OECD, Paris

NEA (2020) Management and disposal of high-level radioactive waste: global progress and solutions. OECD, Paris. https://www.oecd-nea.org/jcms/pl_32567

NEA/IAEA/CEC (1991) Disposal of radioactive waste: can long-term safety be evaluated? An international collective opinion. OECD, Paris

OSPAR (2007) Decision 2007/1 to prohibit the storage of carbon dioxide streams in the water column or on the sea-bed. https://www.ospar.org

OSPAR Convention for the protection of the marine environment of the North-East Atlantic. https://www.ospar.org/convention/text

Savage D et al. (2004) A generic FEP database for the assessment of long-term performance and safety of geological storage of CO_2. Version 1.0. Quintessa, Henley-on-Thames, Oxfordshire

Conventional Toxic Waste

Baccini P, Brunner PH (2012) Metabolism of the anthroposphere. Analysis, evaluation, design. MIT Press, Cambridge MA. https://mitpress.mit.edu/9780262016650/metabolism-of-the-anthroposphere/

FADWO, Swiss Federal Ordinance on the Avoidance and the Disposal of Waste as of 2015-12-4. Status as of 2020-4-1. SR 814.600 (English version with no legal force)

FEPA, Swiss Federal Environmental Protection Act as of 1983-10-7. Status as of 2020-7-1. SR 814.01 (English version with no legal force)

FOEN, Swiss Federal Office for the Environment, BUWAL (2003) Die Sackgebühr aus Sicht der Bevölkerung und der Gemeinden. Schriftenreihe Umwelt, Nr. 357. BUWAL/FOEN, Bern

FOEN/BAFU (2012) Magazine "environment" 2/2012. Kostbare Umweltinfrastruktur [Precious infrastructure for the environment]. https://www.bafu.admin.ch/bafu/de/home/dokumentation/magazin/archiv-magazin–umwelt-.html

FOEN/BAFU (2019) Special waste statistics 2018. https://www.bafu.admin.ch/bafu/en/home/topics/waste/state/data.html

FRPA, Swiss Federal Radiation Protection Act as of 1991-3-22. Put into force on 1994-10-1. Status as of 2017-5-1. SR 814.50 (English version with no legal force)

FWC, Eidg. Kommission für Abfallwirtschaft/Federal Waste Commission (1986) Leitbild für die schweizerische Abfallwirtschaft [Guiding principles for the Swiss waste economy]. Schriftenreihe Umweltschutz, 51. Federal Office for the Environment, Bern

Grunwald A (2016) Nachhaltigkeit verstehen. Arbeiten an der Bedeutung nachhaltiger Entwicklung. oekom, München

Herrmann A, Röthemeyer H (1998) Langfristig sichere Deponien. Situation, Grundlagen, Realisierung. Springer, Berlin

Minamata Convention (2017, on mercury). https://www.mercuryconvention.org/en

US RCRA, Resource Conservation and Recovery Act as of 1976-10-21. Pub. L. 94–580. https://www.govtrack.us/congress/bills/94/s2150/text

Thompson M (1979, [2]2017) Rubbish theory. The creation and destruction of value. Pluto Press, London

Wollmann H (2015) Die Erbringung öffentlicher und sozialer Dienstleistungen zwischen Kommunen, Staat, Privatem und Öffentlichem Sektor im Wandel und Sog der Leitbilder und Reformschübe [Delivering public and social services in communities, the state, and private and public sectors in the course of time and the wake of concepts and reforms]. In: Döhler M, Franzke J, Wegrich K (eds) Der gut organisierte Staat. Festschrift für Werner Jann zum 65. Geburtstag. Modernisierung des öffentlichen Sektors, Sonderband 45. Nomos, Baden-Baden, Germany, pp 531–558 (waste management and policies in France, Germany, Sweden, and UK)

CCS

Archer D (2005) Fate of fossil fuel CO_2 in geologic time. J Geophys Res 110(C09S05):1–6. https://doi.org/10.1029/2004JC002625

Bachu S (2000) Sequestration of CO_2 in geological media: criteria and approach for site selection in response to climate change. Energy Convers Manage 41:953–970

Benson SM (2005) Lessons learned from industrial and natural analogs for health, safety and environmental risk assessment for geologic storage of carbon dioxide. In: Thomas DC, Benson SM (eds) Carbon dioxide capture for storage in deep geologic formations—results from the CO_2 capture project, vol 2. Elsevier, Amsterdam, pp 1133–1141. https://www.sciencedirect.com/science/article/pii/B9780080445700501549

Gale J, Bachu S, Bolland O, Xue Z (2007) To store or not to store? Int J Greenhouse Gas Control 1(1):1. https://doi.org/10.1016/S1750-5836(07)00037-0

GCCSI, Global CCS Institute (2020) The value of carbon capture and storage. GCCSI, Melbourne, p 23

Gomes GN, Corazza RI (2019) Technological knowledge production towards climate change mitigation. Conference paper at IV ENEI Encontro nacional de economia industrial e inovação. São Paolo, 10–12 Sept 2019, p 20

Grünwald R (2008) Treibhausgas – ab in die Versenkung? Möglichkeiten und Risiken der Abscheidung und Lagerung von CO_2. Studien des

Büros für Technikfolgen-Abschätzung beim Deutschen Bundestag. Global zukunftsfähige Entwicklung, vol 25. Edition Sigma, Berlin

IEA, International Energy Agency (2020). The role of CCUS in low-carbon power systems. https://www.iea.org/reports/the-role-of-ccus-in-low-carbon-power-systems

Jacobson MZ (2009) Review of solutions to global warming, air pollution, and energy security. Energy Environ Sci 2:148–173

Leung DYC, Caramanna G, Maroto-Valer MM (2014) An overview of current status of carbon dioxide capture and storage technologies. Renew Sustain Energy Rev 39:426–443

Li Q, Liu G (2016) Risk assessment of the geological storage of CO_2: a review. In: Vishal V, Singh TN (eds) Geologic carbon sequestration. Springer, Cham, Switzerland, pp 249–284. https://www.springerprofessional.de/en/risk-assessment-of-the-geological-storage-of-co2-a-review/10098344

NOAA, US National Oceanic and Atmospheric Administration (2020) Climate change: atmospheric carbon dioxide. https://www.climate.gov/, https://www.climate.gov/news-features/understanding-climate/climate-change-atmospheric-carbon-dioxide#:~:text=But%20increases%20in%20greenhouse%20gases,Earth's%20long%2Dlived%20greenhouse%20gases

Romasheva N, Ilinova A (2019) CCS projects: how regulatory framework influences their deployment. Resources 8(181):19. https://doi.org/10.3390/resources8040181

Zappone A et al (2021) Fault sealing and caprock integrity for CO_2 storage: an in-situ injection experiment. Solid Earth Discuss, EGU, https://doi.org/10.5194/se-2020-100. Preprint

Websites

https://www.oecd-nea.org/ (Nuclear Energy Agency of the OECD: radioactive waste)

https://www.iaea.org/ (International Atomic Energy Agency: radioactive waste)

https://www.eea.europa.eu/themes/waste (European Environment Agency: conventional waste)

https://www.iea.org/ (International Energy Agency: carbon storage)

https://www.ipcc.ch/ (Intergovernmental Panel on Climate Change: carbon storage)

https://www.globalccsinstitute.com (Global CCS Institute: carbon capture and storage)

http://www.cgseurope.net (Pan-European coordination action on CO_2 geological storage) (also Chaps. 5 and 7)

Sources

Grunwald A (2016) Nachhaltigkeit verstehen. Arbeiten an der Bedeutung nachhaltiger Entwicklung. oekom, München (Box 4.1)

IPCC, Intergovernmental Panel on Climate Change (2005) IPCC special report on carbon dioxide capture and storage. Metz B, Davidson O, de Coninck HC, Loos M, Meyer LA (Working Group III). Cambridge University Press, Cambridge (Fig. 4.7)

Nagra (2014) Modellhaftes Inventar für radioaktive Materialien MIRAM 14. NTB 14–04. Nagra, Wettingen, Switzerland (Fig. 4.4)

Projectdataresearch (2022, web) Project & bid information for nationwide lining, landfill, excavating & mining projects. http://www.projectdataresearch.com/landfills.html (Fig. 4.6)

SKB, Svensk Kärnbränslehantering AB, Swedish radioactive waste implementer (2022, web) https://www.skb.com/future-projects/the-spent-fuel-repository/our-methodology/ (Fig. 4.5)

Solomon S, Plattner GK, Knutti R, Friedlingstein P (10 Feb 2009). Irreversible climate change due to carbon dioxide emissions. Proc Nat Academies Sci, PNAS 106(6):1704–1709 (incl. supporting information: pnas.0812721106) (Fig. 4.3)

UNEP, United Nations Environment Programme et al (2004) Vital waste graphics. Basel Convention, Grid Arendal, UNEP, DEWA Europe. https://www.grida.no/publications/264 (Fig. 4.1)

Systems, Governance and Institutions

5

Abstract

The standard (technical) robustness concept (of Chap. 4) is enlarged by societal and institutional, including regulatory, aspects, and then amplified by the more dynamic notion of "resilience" and "adaptiveness", followed by an analysis of processes and procedures in all three policy areas under scrutiny: radioactive waste, conventional hazardous waste and carbon storage CCS. The approach culminates in turning over the concept of "governance" to the applications, thus establishing a framework for the intended "Strategic Monitoring".

Keywords

System • Institution • Organisation • Time, long-term • Robustness • Resilience • Adaptiveness/adaptability • Governance • Evaluation • SWOT method • Radioactive/nuclear waste • Conventional toxic/hazardous waste • Carbon Capture, Utilisation and Storage, CCUS/CCS

Learning Objectives

- Expand your view of robustness to resilience and governance;
- Acknowledge different views on what is "long-term";
- Accept the "trans" character of the issues: transdisciplinary, transscientific, transgenerational, transpolitical;
- Learn relevant details about the three-step approach.

5.1 From Robust Technical Systems to Integral Robustness

As pointed out before, science cannot define what waste[1] is, in either field. Procedural and process issues make it explicit that perspectives (and their changes) and dynamics play an important role in the management of nuclear waste, conventional waste and CO_2 in the long run (Krütli et al. 2012). This argues that the standard concept of (technical) "robustness" is necessary to ensure safety but not sufficient.

Risk attributes not dealt with in risk studies play a vital role in the public dispute (Wynne 1980; Hansson 1989), sometimes not prone to be covered by experts due to "overcomplexity". In nuclear technology, this may be the connection of civil and military use or proliferation, the "normality" of disasters with system immanent failure (Perrow 1982, 1984) or the longevity and irreversibility of potential impacts. According to Wynne, the public appraises a technology as a whole including its institutions (Wynne 1980).

The systems approach (Box 5.1) spells out the fact that the various project phases of a facility site require at least several decades, from characterisation, design via operation to monitoring and closure/sealing (Fig. 3.2). Consequently, management fundamentals and concepts are liable to an integration into a consistent and stepwise decision-making process. Their instruments have to be designed in a dynamic, adaptive, even experimental manner (Cook et al. 1990; Ascher 1999, 375), but the ultimate goal remains, viz., the passive protection of present and future generations and environments. According to Rip, a system is "socially robust" if most arguments, evidence, social alignments, interests and cultural values lead to a consistent option (Rip 1987, 359). Therefore, the concerned and deciding stakeholders have to eventually achieve consent on some common interests, along the lines suggested in Table 3.1. According to Bijker and colleagues, **"Closure"** in science occurs "when a consensus emerges that the 'truth' has been winnowed from the various interpretations" (Bijker et al. 1987, 12).

---

[1] For key terms and concepts, refer to Glossary in the back.

T. Flüeler, *Governance of Radioactive Waste, Special Waste and Carbon Storage*, Springer Textbooks in Earth Sciences, Geography and Environment, https://doi.org/10.1007/978-3-031-03902-7_5

Box 5.1: System

"A system consists of subunits with certain boundary conditions between which … processes take place", "organized complexity … with strong interactions", "set of elements standing in interrelations", "the basis of the open-system model is the dynamic interaction of its components." (von Bertalanffy 1968, 21, 19, 55, 150)

Organisations, Part 1

"Characteristic of organization, whether a living organism or a society, are notions like those of wholeness, growth, differentiation, hierarchical order, dominance, control, competition, etc." (von Bertalanffy 1968, 47).

Structure

"The arrangement and organization of mutually connected and dependent elements in a system or construct" (Oxford English Dictionary, entry 3a).

"A combination or network of mutually connected and dependent parts or elements; an organized body or system" (Oxford English Dictionary, entry 7).

	Nagra estd.1972	1980	1990	2000
Separation promotion/oversight	Pa	IAEA	IAEA	(✓)
Publication	P E	Pa ✓		
Traceability	E E	P	✓	✓
Criteria	E E			✓
Independent reviewing	P E ✓		P	✓
Host rock		E ✓		
Controllability	E P			✓
Retrievability		E P		✓
Disposition concept	**D** **FD**		✓	**eFD**
Transparent funding			Pa/P	✓
Public involvement	P	P (✓)P	P (v)	P (v)

Fig. 5.1 Integration of relevant aspects into the official disposition concept for radioactive waste in Switzerland. Note the time difference between first issue raising (P, E, Pa) and its adoption in the system/programme (✓). The empirical basis for the table is the content analysis over 40 years by Flüeler (2002). Pa Parliament, E Expert(s), P Public (v: vote at potential repository site Wellenberg 1995/2002), IAEA International Atomic Energy Agency (recommendation; Waste Convention 1997), D underground disposition, FD final disposal, eFD extended final disposal/monitored long-term geological disposal (acc. to EKRA 2000) (reproduced and translated from Flüeler 2002d, 170)

"**Overall robustness**", in a way, is a fuzzy notion, but it recognises the complex sociotechnical character of the issues and has the potential to stepwise and iteratively integrate structural and procedural/dynamic elements—as well as various and diverse types of uncertainty—into long-time governance. Over time, various stakeholder groups have raised critical and crucial topics which, eventually, were adopted by the institutions in charge (Fig. 5.1). These are, specifically in Switzerland, the following: separation of promotion and oversight in the Federal administration, adequate funding according to the causality principle, extensive duty of publication, traceability of reasoning, transparent formulation of criteria (for siting, inventory, etc.), controllability, retrievability, extensive independent, i.e., pluralistic, reviewing, stepwise and phased procedure and participation of the public (e.g., also in oversight committees).

The system calls for multiple technical barriers against the release of radioactivity or other toxic substances, as well as for phased societal checks to achieve and sustain confidence in technical assessments and, hence, acceptance or at least tolerability in society (Fig. 5.2). This implies that many institutions have a role to play with various functions (e.g., Flüeler 2000b, c, 2006b, 2014b, 2015, 2021; Flüeler and Scholz 2004b; Kuppler and Hocke 2019; Metlay 2021), in a communicative atmosphere of mutual respect and learning (Chap. 6).

The primary goal in radioactive waste governance is the stability of the system: the permanent protection of humans and the environment from the release of (harmful) radioactivity. The complimentary goal is flexibility, defined as the potential to intervene. A conclusive programme for control and monitoring has to be specified, including publication of the work, intensive reviewing, respective quality assurance and wide involvement of affected and concerned parties. In compliance with the International Waste Convention (IAEA 1997), the nuclear waste issue is considered a national task, which has to be carried out on the territory of the waste-producing country and on the basis of current knowledge, as a Swedish advisory body suggested (KASAM 1998, 4).

We are faced with a pronouncedly long-term project: "A repository is, by definition, a long-term project, extending over centuries … or even much longer periods … involves a relatively long lead time (possibly more than 20 years [sic!] for [high-level waste] or spent fuel) and is then anticipated to receive waste during several decades. After closing the repository, a surveillance and monitoring period will almost certainly be carried out even [sic!] for shallow land burial type repositories with [low- and intermediate-level waste]. This underlines once again the importance of the continuity factor not only from a contractual, but also from a technical,

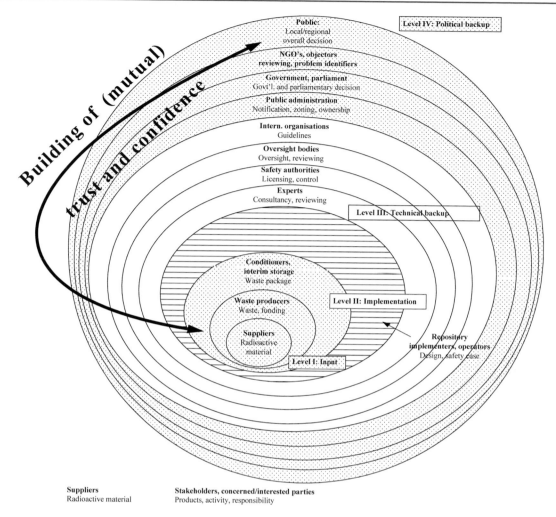

Fig. 5.2 Societal/institutional robustness. Stakeholders are to act according to their respective responsibilities. Dependent on their mutual trust, their activities serve as institutional barriers and potentially lead to a consistent, i.e., robust decision, backed up by incremental building of confidence in the overall disposal system and trust in the respective actors. System levels are indicated (reproduced from Flüeler 2002d, 2004a, 2006b, 259)

point of view (possibility/obligation to transfer/receive waste, waste acceptance criteria and quality of waste, control and monitoring, etc.)" (IAEA 1998, 9). Consequently, it is of utmost relevance that the various "barriers" adequately mesh. Figure 5.3 depicts the integration of technical and societal/institutional aspects into overall— or **"integral"**— robustness (Fig. 5.3).

In order to achieve this aim, it is crucial that long-term comprehensive, but also stepwise, planning is set up. If the concerned public indeed is given an adequate share, the chances rise that the decisions taken will also be accepted in the future as being legitimate (Espejo and Gill 1998).

5.2 What is "Long-Term", Really? Definitions and Implications

We learned that there is no objective definition of "waste". When coming back to the time dimensions, we will see that this also applies to what "long-term" is. What it means depends on the chosen perspectives—who defines, what the context is and what it is defined for. In actual fact, we talk about different views on what time frames to look at (from Flüeler 2006d, expanded).

We may distinguish among the following sensibilities[2]:

---

[2] Knowing that time is a multifaceted concept, especially in our fields (cf. Moser et al. 2012a, b). The starting point of the arrows is "today".

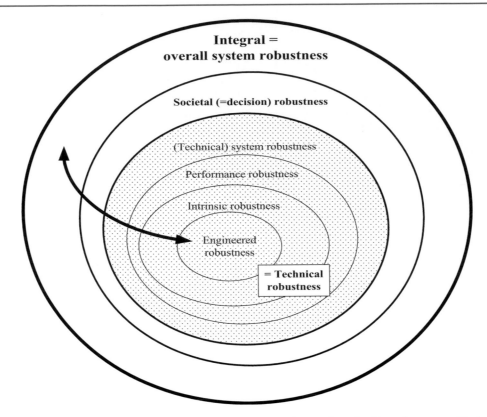

Fig. 5.3 Integral robustness. Overall system robustness depends on (dotted) technical robustness (Sect. 4.6) and societal (=decision) robustness. The sequenced symbolic portrayal of the types of robustness shall not infer an objectivistic approach. The final and decisive validation of the concept is the implementation of a disposal concept with demonstrated long-term safety backed up by respective decisions and actions. This presupposes an effective coupling, i.e., complementing/meshing, of the various types of robustness (reproduced from Flüeler 2001a, 319 and 2006b, 265)

Technical (Waste Disposal) Perspective

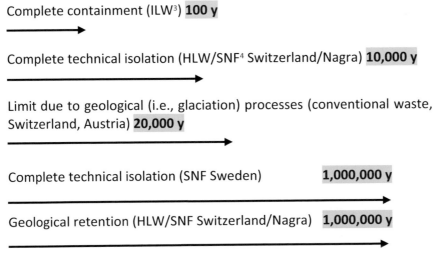

Complete containment (ILW[3]) **100 y**

Complete technical isolation (HLW/SNF[4] Switzerland/Nagra) **10,000 y**

Limit due to geological (i.e., glaciation) processes (conventional waste, Switzerland, Austria) **20,000 y**

Complete technical isolation (SNF Sweden)   **1,000,000 y**

Geological retention (HLW/SNF Switzerland/Nagra) **1,000,000 y**

Comment: Depending on waste characteristics, disposal objectives and disposal design even technical time frames span over orders of magnitude. It is implied that "technical" also means environmental since the protection of the (long-term) environment is aimed at.

---

[3] ILW: Intermediate-level waste.
[4] HLW: High-level waste, SNF: Spent nuclear fuel.

Technology (policy) perspective

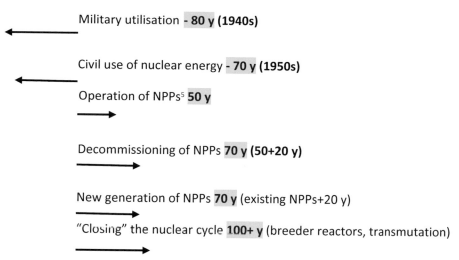

Comment: Energy policy plays an important (and driving) part in the (nuclear) issue. Nuclear history has to be taken into account. Whether or not the nuclear path is perpetuated, "long-term" assumes another slant and responsibility is passed on to a greater or lesser degree to future generations (continuation or phase-out, respectively).

Institutional Perspective

Short-lived waste[6] **30 y**

→

Operation of waste storage[7] **50 y**

→

Operation of waste disposal[8] **60-70 y**

→

Post-closure of waste disposal: active safety measures[9] **+70-300-500 y**

→ – ▶····▶

Post-closure of waste disposal: passive safety measures[10] **+300-+500 y**

→ – –▶·····▶

---

[5] NPP: Nuclear power plant.
[6] As defined by the technical nuclear community (radionuclide half-life period less than 30 years, IAEA 1994).
[7] Storage denotes (intermediary and) controlled disposition of waste (with the intention of retrieval).
[8] Disposal denotes (final) and, ultimately, uncontrolled disposition of waste (without the intention of retrieval).
[9] *in situ* monitoring, environmental surveillance, etc.
[10] Land-use restrictions, land register.

Post-closure of waste disposal: memory **+500 y**

⟶

Comment: What is "short-lived" was defined by a technical nuclear institution (International Atomic Energy Agency, IAEA 1994). There are continuous transitions between different phases, again depending on waste
characteristics, disposal objectives and disposal design. Specifications, e.g.,
for monitoring, have to be given with regard to purpose (why, what for),
locations (where), parameters (what) and methods (how).

Societal Perspective

One generation **30 y**

⟶

Working population (responsible for business decisions, generation[s]
"n-1, n") **50 – 60 y**

⟶

Contemporaries
(filial generation, parental generation II [working population], parental
generation I [pensioners]) **100 y**
(n+1)   (n-1, n) (n-2)

⟶

Yardstick of the "seven generations"
(Canadian First Nations' target for assessment or evaluation of consequences of current issues[11]) **175 y**

⟶

Comment: In general, societal perspectives depend on the societies they
apply to—contemporary Western European notions evidently are different
from Canadian First Nations or from Chinese or from Middle Age European
notions. The history of nuclear activities denotes that the n-2/n-3 generation (-70 to -80 y) started the nuclear enterprise. The waste programme set-
backs and the waste longevity make the issue a "long-term" problem and
require a transfer to future generations (Fig. 3.2): a transfer of disposition
rationale (and options), of know-how, of resources, of procedure and of
institutional constancy.

---

[11] "Traditionally, no decision was made until it was understood how it
would affect the next seven generations" (https://healingoftheseven
generations.ca/about/history/).

Political Perspective

Term of office (**politics**), legislation **4 – 5 y**
→

Scope of the radioactive waste management **policy 300 – 500 y**
——————————→

Guardian of the process (programme policy) (e.g., 10x term of office) **40 – 50 y**
——————→······►

Comment: Decade-long programmes are prone to political volatility and arbitrariness: <u>N</u>ot <u>I</u>n <u>M</u>y <u>T</u>erm <u>O</u>f <u>O</u>ffice (NIMTOO). Consequently, it suggests itself to establish some sort of a "guardian" of the radioactive waste management policy to overcome discretionary politics and to see to it that the programme is on target. In view of the "trans" character of the issue (beyond party politics: "transpolitical", and more than an interdisciplinary scientific issue: "transscientific"), it is suggested that the body be pluralistically composed of knowledgeable, trustworthy personalities, highly respected by society and not driven by daily politics. This demonstrates the interconnection of institutional, societal and political perspectives (see Chaps. 4 and 7).

Individual Perspective

Attitude and worldview **0 – n y**
►·····················►

Comment: Individuals' worldviews are very heterogeneous ("n" can be any number). Usually one encounters a preference for the present, at least a high discount rate, i.e., time periods beyond close future (and thus future generations) are not relevant to the majority of people.

Economic Perspective

—————►
Quarterly to decennial **0.25 – 30 y**

Comment: The perception of time frames in companies ranges from three-month calculations (and related pay-back periods) to 30-year scenarios as a basis for investment decisions. Economic growth (and thus decisive interest rates) cannot be calculated beyond 3 to 5 years.

Time—Past, Present, Future

What now is time? If nobody asks me I know;
but if I wanted to explain it to somebody on his quest,
I do not know. This I can say with confidence:
I know that there would not be passed time
if nothing went by,
and no future time if nothing were present.

Augustinus Aurelius, 354-430, Bishop of Hippo,
philosopher, Confessiones, 397-401, Book XI, Chapter 14

… not the times are malicious but our actions.
And we are time.

Augustinus Aurelius,
In epistulam Ioannis ad Parthos, tractatus IX, 9

You are time. If you are good, the times are good.
Augustinus Aurelius

Le temps te construit des racines.
Time builds roots for you.

Antoine de Saint-Exupéry (1900–1944), Citadelle, 1948

Die Vergangenheit hat mich gedichtet/
ich habe/die Zukunft geerbt/
Mein Atem heisst/jetzt.
The past has versified me/
I have/inherited the future/
My breath is/now

Rose Ausländer (1901–1988), Mein Atem (excerpt), 1978

Apart from the technical challenges, the main societal and institutional challenge is to match and reconcile the geologic long-term dimension of (passive) protection (Fig. 5.4a, b) with the longevity of a coherent programme and its targeted implementation (Fig. 5.4c, d).

5.3 From Robustness to Resilience

The overall system robustness strived after, across all technical and non-technical subsystems, is more than "engineering resilience", the speed of return to a stable steady state following some perturbation (Pimm 1984). It might amount to what Gunderson and colleagues called "ecological resilience", i.e., the capacity with which a system may absorb disturbance *before* having to be restructured (Gunderson et al. 2002, 4ff; Holling 1996) (Box 5.2). It is not enough to keep a system, a building, an institution, etc., stable or "robust" or, for that matter, from just falling apart. Emergency planning is needed, but if this is meant to create resilience, it is putting the cart before the horse. Resilience is essentially proactive and must be strived for at all levels: the micro (individual and staff), meso (group,

institution and company) and macro (nation-state, inter- and supranational, see McCarthy et al. 2017).

The concept of "**long-term stewardship**" (Tonn 2001; LaPorte 2004) as proposed and practised by the US Department of Energy with respect to legacy radioactive waste (DOE 1999, 2001a, b) falls short of a comprehensive understanding of the issue. Even official analyses of institutional monitoring of US sites demonstrate in frustrating openness: "It is now becoming clear that relatively few … DOE waste sites will be cleaned up to the point where they can be released for unrestricted use. 'Long-term stewardship' (activities to protect human health and the environment from hazards that may remain at its sites after cessation of the remediation) will be required for over 100 of the 144 waste sites under DOE control …. The details of long-term stewardship planning are yet to be specified, the adequacy of funding is not assured, and there is no convincing evidence that institutional controls and other stewardship measures are reliable over the long-term" (NAS 2000, 2).[12] Strohl was of the opinion in 1995 already: "… institutional instruments, although indispensable with regard to long-term safety, should only be considered as making a contribution of relative importance and of limited duration, and this must be made clear" (Strohl 1995, 25f). This perspective has been maintained ever since (NEA 2014a).

It was, and still is, widely recognised that (high-level) nuclear repositories are "first-of-a-kind, complex and long-term projects that must actively manage hazardous material for many decades" (NRC 2002, 1). Thus, adequate measures must be taken. The institution(s) in charge of and entrusted with stewardship must be more than a "guardian" who does the "stopping activities that could be dangerous" (NAS 2003, 2), more than a "watchman", "land manager", "repairer", "archivist", "educator to affected communities" or "trustee … assuring the financial wherewithal" (ibid., 2). It is not sufficient to guarantee constancy; the guardian must be respected and trusted by the majority of the other parties involved. In many countries, nuclear-centred state agencies have a mixed reputation (Torfing 2006; Probst and McGovern 1998).

Kuppler and Hocke ask "which institution will care for the repository over decades or eventually longer periods like some centuries?" (Kuppler and Hocke 2019, 1352). Even though the author has long voted for a "guardian" body—and still does—a thorough system analysis suggests that there will be no one institution to do that. The entire process along the proposed three steps (Chap. 3) will have to be both tight and flexible to secure a "rolling present" with an ongoing targeted, yet adaptive, programme.

---

[12] This still holds true over the years (DOE 2012; NCSL 2017, 2022).

a Potential Swiss siting region North of Lägern – Wehntal (20 km northwest of Zurich), today

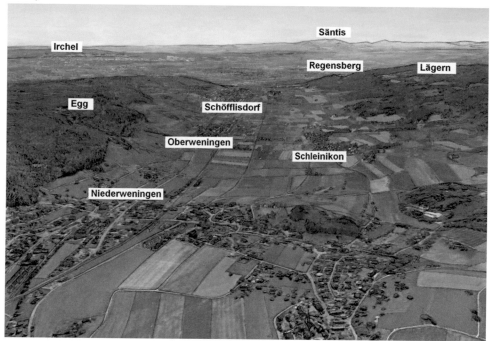

b Potential Swiss siting region North of Lägern – Wehntal, 140,000 years before present

Fig. 5.4 Different dimensions of what "long-term" means: objective (**a**, **b**), institutional (**c**, **d**) (reproduced from: **a b** Mammutmuseum.ch, **c** Fuchs 2016, **d** US DoD)

c Today: Swiss site-selection procedure, (final) phase 3 started – Dome of Cologne, 2016

d Tomorrow: 70 yrs from now, waste facility closed – Dome of Cologne, around 70 yrs back: April 1945

Fig. 5.4 (continued)

Box 5.2: Resilience

– Ecological: "the capacity of a system to absorb disturbance and reorganize while undergoing change so as to still retain essentially the same function, structure, identity, and feedbacks" (Walker et al. 2004);
– Organisational: human "resilience is developed, sustained, and grown through discourse, interaction, and material considerations. There are several communicative processes involved. These include: (a) crafting normalcy, (b) affirming identity anchors, (c) maintaining and using communication networks, (d) putting alternative logics to work and (e) downplaying negative feelings while foregrounding positive emotions" (Buzzanell 2010).

Adaptability

– Adaptive systems: "Complex adaptive systems involve many components that adapt or learn as they interact" (Holland 2006);
– Adaptability: ability to change something or oneself to fit to occurring changes, with or without an external demand for change (Andresen and Gronau 2005).

5.4 Excursus: COVID-19—Lessons from a Virus to Humanity

Even though pandemics and wastes are two different issues, some insights might be learned from the ongoing spread of the SARS-CoV-19 virus all over the globe and the way we treat it.[13] China informed the world on New Year's Eve 2019 about a new virus discovered in the fish market of the City of Wuhan; less than two and a half months later, on 12 March 2020, the World Health Organization, WHO, declared the outbreak of a pandemic.[14] It had and still has enormous impacts on all aspects of life: individual, social, economic lives, even on the civil society and politics. Some characteristics may be highlighted:

- Disease uncertainties: Transmission paths, emergence of new variants, incubation time, often asymptomatic appearance, gender and age differences, tracing methodology, etc.—there was a long-standing dispute among experts;
- Response types and times: Some countries decreed a rapid and strict lockdown at an early stage (e.g., South Korea and Victoria in Australia), whereas others counted on herd immunity (the United Kingdom for some time) or slackened the reins until 2021 (USA);
- Response philosophies: Trust in players and systems was very diverging (self-responsibility in Sweden vs. curfews in Singapore);
- Systemic risks: The health systems with their infrastructure (hospitals, intensive care units: staff, beds) were stretched to and beyond their limits; they were clearly defined as systematically relevant (just like information and communication technology or energy supply—too important to fail);
- Use of language: Technocratic terms like "social distancing" were deconstructed ("come close but keep distance!");
- Perception of risk and measures depend on perspectives: innkeepers, hairdressers, elderly people, pupils, teachers, bus drivers, workers, etc.;
- Not intended side effects: The lockdown of educational systems led to a reduction of performance and intensification of social inequalities as the access to IT varied, especially for educationally disadvantaged; challenges for all: digital tools, distance learning, novel socialising …;
- Trade-offs in ethics: Individual liberties versus control of disease, triage of patients, segregation of risk groups, disadvantages of economic groups, etc.

Coronavirus disease exposed structural inequalities within and between states and vulnerabilities in states with weak or ill-prepared health systems which had not been qualified as such before. According to the Global Health Index, "[n]ational health security is fundamentally weak around the world. No country is fully prepared for epidemics or pandemics, and every country has important gaps to address. Countries are not prepared for a globally catastrophic biological event" (GHI 2019). It is interesting to see the differences between their preparedness *ex ante* judged by experts and their actual performance during the ongoing COVID-19 pandemic (Table 5.1). Experts' views (on national health systems) do not pass the reality check, the ongoing COVID pandemic, where these systems could have been validated. (Some of) the best ranking countries score badly (the US, the UK and the NL), whereas others have done relatively well during the pandemic (Vietnam, China, Taiwan; as to available information). For instance, "public health vulnerabilities" as an indicator group is meant to reflect the "ability of a

---

[13] Recognising that this is very daring as a Google search alone gave 473 mio entries for "COVID-19 lessons learned" in autumn 2020 (2020–09–16, 13:00 CEST). At the beginning of 2022, the figure was 652 mio entries (2022–01–26, 17:15 CEST).

[14] https://www.who.int/emergencies/diseases/novel-coronavirus-2019.

Table 5.1 Global Health Index, GHI 2019 versus experience during the Covid-19 pandemic. Explanations are given in the text. Snapshot as of 16 September 2020 (Sources:*)

Rank	GHI (2019) (195 countries)	Total of infected persons (mio)	Infections per million inhabitants	*Least* infections per mio inh. (selection)	Deaths per mio inhabitants	*Least* deaths per mio inhabitants (selection)
*	a	b	c, d	c, d, e	d, f	d, e, f
1	USA (83.5 out of 100)	USA 6.61	Chile 22,912	Laos 3	Peru 938	Cambodia/Eritrea/Laos 0.0
2	United Kingdom UK	India 5.02	Peru 22,383	Tansania 9	Belgium 857	Taiwan 0.29
3	Netherlands NL	Brazil 4.38	Brazil 20,617	Vietnam 11	Spain 642	Tanzania 0.35
4	Australia	Russia 1.07	USA 19,958	Cambodia 16	Bolivia 638	Vietnam 0.36
5	Canada	Peru 0.74	Israel 19,270	Taiwan 21	Chile 630	Sri Lanka 0.61
6	Thailand	Colombia 0.73	Colombia 14,319	Niger 49	Brazil 626	Myanmar 0.72
7	Sweden	Mexico 0.68	Argentina 12,511	Thailand 50	Ecuador 621	Thailand 0.83
8	Denmark	South Africa 0.65	Costa Rica 11,413	China 63	UK 614	Mozambique 1.18
9	South Korea	Spain 0.60	Bolivia 10,990	Chad 66	USA 592	Uganda 1.27
10	Finland (68.7)	Argentina 0.57	South Africa 10,985	Yemen 68	Italy 589	Rwanda 1.70
	Vietnam (50.) 49.1 China (51.) 48.2 Taiwan: not analysed	Chile (11.) 0.44 Iran (12.) 0.41 France (13.) 0.40 UK (14.) 0.37 Italy (20.) 0.29 Canada (26.) 0.14 China (31.) 0.090 Sweden (32.) 0.087 NL (33.) 0.085 Australia 0.03 Thailand 0.003	Singapore (11.) 9826 Sweden (14.) 8649 NL 4943 UK 5513 Canada 3703 India 3638 Denmark 3475	South Korea 439 Australia 1049 Finland 1575	Sweden (11.) 579 NL (16.) 265 Canada (21.) 243	China (15.) 3.29 South Korea (33.) 7.16 Australia 32 Finland 61 Denmark 109

a https://www.ghsindex.org/archive/
b https://ourworldindata.org/coronavirus, https://github.com/owid/covid-19-data/tree/master/public/data
c https://ourworldindata.org/grapher/total-confirmed-cases-of-covid-19-per-million-people?tab=table&time=2020-01-22..latest
d total confirmed cases, countries with population of >5 mio
e actual rank in brackets, GHI countries of column 2 included
f https://coronavirus.jhu.edu/data/mortality (accessed 16 September 2020)
Acknowledging that there is a dispute on whether a person died <u>of</u> SARS-CoV-19 or <u>with</u> it

country to prevent, detect, or respond to an epidemic or pandemic", i.a., the healthcare access and quality (GHI 2019, 65). The partial score of 93.3/100 for the USA does not match with the COVID-19 experience so far; the death toll was over 260,000 in November 2020, over one million in September 2022.[15] Or Switzerland, ranked 13 in the overall score (67/100), indeed had a pandemic concept (of 2017) which required the stock of masks. At the beginning of the COVID-19 pandemic, in March 2020, the federal administrators argued that wearing masks would not be useful— reports showed that this assertion was based on the fact that masks were not available to the public (TA 2020).

Overall, not the management's thinking of prohibition and command-and-control proved to be successful but the firm and trustworthy demeanour of politicians and communicators, setting clear goals,[16] abiding by their own rules, based on sound science with scientific advisory bodies supporting public health and governance from the very beginning. This made it possible for people to accept the recommended measures, gain skill in a new normality (hygiene rules adhered to, spatial distancing and mask wearing in crowds and enclosed spaces) and even tolerate

[15] Worldometer: 262,701 cases for 23 November 2020, over 1,113,808 for 27 January 2023 (https://www.worldometers.info/coronavirus).

[16] Slow down spreading of virus, cure infected persons, minimise health risks, have as many vaccinated as possible, protect health and minimise negative impacts on society and economy (knowing that there are contradictions between solidarity and necessary trade-offs).

some infringements on personal liberties. This all was in view of the common good of solidarity, "the most important resource in the fight against COVID-19", as WHO's Director-General exclaimed in April 2020 (WHO 2020a): "Health is not a cost, it's an investment" (WHO 2020b). Note that Gross Stein had long ago, in her analysis of the "cult of efficiency" in Canada, identified the health system to be systematically relevant, incidentally together with the educational system (Gross Stein 2002)—talking about literacy (Sect. 7.1).

Facts Instead of Fear, Limit Spread of Virus and Spread of Misinformation

This outbreak is a test of solidarity—political, financial and scientific. We need to come together to fight a common enemy that does not respect borders, ensure that we have the resources necessary to bring this outbreak to an end and bring our best science to the forefront to find shared answers to shared problems. Research is an integral part of the outbreak response. The only way we will defeat this outbreak is for all countries to work together in the spirit of cooperation. This is the time for facts, not fear. This is the time for science, not rumors. This is the time, not stigma.

AG Tedros, WHO Director-General, from Smidt 2020

Now is the time to use the tools we have: our best science, our global ability to share information and our collective cooperation under the UN system to ensure we can contain these outbreaks. If we let misinformation and panic take hold, we will limit our ability to create a global solution for what could potentially grow to a global epidemic.

Smidt 2020

5.5 Governance

How does one set up a programme which has the capability to follow and reach goals once agreed upon but also the flexibility to allow changes where necessary over the time span of decades? The notion of "**Governance**" comes in (Box 5.3).

Box 5.3: Governance

"Governance is about the rules of collective decision making in settings where there are a plurality of actors or organisations and where no formal control system can dictate the terms of relationship between these actors and organisations" (Chhotray and Stoker 2009, 3), with four key elements:

Rules: Incorporate established institutional forms and procedures (legislation, guidelines, licensing, etc.) as well

as "rules-in-use" (Ostrom 1999, 38) where informal conventions and participatory elements come into play.

Collective refers to the fact that a multitude of actors, institutional and individual, are on the stage, with issues of mutual influence and control. A substantial "stakeholder involvement" includes rights for (more) actors to have a say but, conversely, also their responsibility to accept collective decisions.

Decision making: Deciding is not just the preference of an option. Decision making includes the process of deciding, the judgement made, the choice taken and, ideally, the decision implemented (Flüeler 2006b, 101ff). This can be on various dimensions and levels (global to local, strategic to day-to-day business).

No formal control system: There is no "governor in charge" of all, decisions are not top down but inclusive, they rely on communication, deliberation, negotiation and mutual influence.

(Nation-state) Governance was also defined as "the set of traditions and institutions by which authority in a country is exercised. The political, economic, and institutional dimensions of governance are captured by six aggregate indicators". These are: voice and accountability, political stability and absence of violence, government effectiveness, regulatory quality, rule of law, control of corruption (IBRD/Worldbank 2006, 3).

Global Governance, today, "refers to *the exercise of authority across national borders as well as consented norms and rules beyond the nation state, both of them justified with reference to common goods or transnational problems*". This acknowledges that international (e.g., European Union) and transnational (e.g., Greenpeace) institutions play a part (Zürn 2018, 3f, Italics in the original).

Institutions and Organisations, Part 2 (cf. Box 5.1)
"Institutions are the formal and informal rules and norms that organise social, political and economic relations." (North 1990)

International institutions: European Union, etc.

Transnational institutions: Greenpeace, Friends of the Earth, etc.

Epistemic institutions: "international institutions involved in creating shared scientific knowledge": International Panel on Climate Change, IPCC; Nuclear Energy Agency, NEA; International Atomic Energy Agency, IAEA; Worldbank, etc. (Meyer 2013).

Organisations are "groups of individuals bound by a common purpose", "shaped by institutions and, in turn, influence how institutions change" (GSDRC 2022, website).

Governance as a problem-solving concept has emerged with increasing globalisation and the cry for more democratisation (Chhotray and Stoker 2009), with a parallel decline and deepening in global governance (Zürn 2018) not investigated in the present analysis. It may be reasoned that with globalised production also waste and environmental stress globalised (see also Lange 2017). This, in turn, has led to resistance—globalised and regionalised at the same time. Both have posed challenges to the basic units of political organisation and economy, respectively: the nation-state and the company. On the positive side, harmonisation of standards and exchange of knowledge and experience also took place. Governance is a practice, of political and human activity, in the context of our bounded rationality. Dunn calls governance the "cunning of unreason" (Dunn 2000).

Beck observed that "social, political, economic and individual risks increasingly tend to escape the institutions for monitoring and protection in industrial society" (Beck 1994, 5f). He proposed "reflexive modernization" as a counterweight to "the effects of risk society that cannot be dealt with and assimilated in the system of industrial society" (ibid.). In research, technology assessment provides an instrument for coping with unintended consequences. In this vein, Voss and colleagues developed the notion of "**reflexive governance**" and took the interconnected issues of complexity, uncertainty, path dependence, ambivalence and distributed control as a starting point to characterise the governance problem surrounding sustainable development. They derived six strategies to prevent societal development from being undermined by the unintended detrimental effects of steering activities: (1) integrated knowledge production, (2) experiments and adaptivity of strategies and institutions, (3) iterative, participatory goal formulation, (4) anticipation of long-term systemic effects of measures (developments), (5) interactive strategy development and (6) creating congruence between problem space and governance (Voss et al. 2006).

5.6 Procedure: Examples

To bring the concept to life, the three-step procedure proposed in Section 3.4 is demonstrated with real approaches in various countries.

Step 1, Discuss: Evolving from a Linear to a More Pluralistic Model

Having learned from the failures at Wellenberg, the planned site for low- and intermediate-level **radioactive waste**, where the cantonal voters rejected an initial as well as a developed project in 1995 and 2002, respectively, the Swiss Confederation started an extensive consultation about a site-selection process to be set up. The underlying concept

document was reviewed in several rounds by technical and political actors; in fact, it was a comprehensive stocktaking in the policy field (BFE 2008, 5), even though it was not as widespread as the Canadian dialogue called "Choosing a way forward" (NWMO 2005, 2022). The ongoing so-called Sectoral plan for deep geological repositories includes a thorough "regional participation" of potential siting regions (Kuppler 2012). When setting up "regional conferences" in each of six geologically suitable areas it was given special attention to consider the entire societal spectrum, from political parties to churches, though females and youngsters were and still are underrepresented (Planval 2014). The conferences have a consultative character; to give them sufficient legitimation (and stability), the quorum of elected members of municipal councils was set to be 50%. Experience so far shows that the safety-first paradigm is accepted on all levels—national, cantonal, regional and communal. Regional stakeholders have been able to have a say, especially on the spatial-planning topics (Fig. 6.4).

Overall, it is essential to allow a comprehensive discourse in society about the matter under scrutiny so that no pertinent issues go by the board. The approaches in various countries evolved over time (Fig. 5.5), basically from technocratic to more pluralistic models as the former failed in every case. Kemp (1992) analysed the nuclear waste decision-making process in seven industrialised nations in a study pioneering regarding systematics but, evidently, outdated with respect to state programmes (Kemp 1992, 167). According to that author, one extreme approach is DAD, "Decide–Announce–Defend". Public involvement is minimised, the authorities in charge have the process under control; therefore, it was labelled "centralised" and "closed". The other extreme is—in ideal terms—"devolved" and "open" planning with the steps: "Establish criteria–Consult–Filter–Decide". The public is involved at an early stage; "filtering", doubtlessly the act of trading off, takes place in a phased and transparent manner.

An early and broad **goal discussion** has to be established to increase the legitimacy of the key conceptual decisions (Klinke and Renn 2001), regarding decision impacts to be borne by future generations (termination of control, uncovering reasons for retrievability, etc.) and for an early detection of "system and decision weaknesses" (no technical or political, safety-compromising, "fixes"); thereby society's risk of enormous costs of corrective actions may be reduced. If appropriately set up, votes may be viewed as inclusive reviewing; as a much more representative instrument, they are not replaceable by "more modern" modes of participation but may be complemented. In fact, they should be, as votes (opposed to elections) are in many countries not part of the political culture and as considerate participation, which better reflects public opinions than plain yes/no decisions. Well-designed participatory tools also structure a discourse

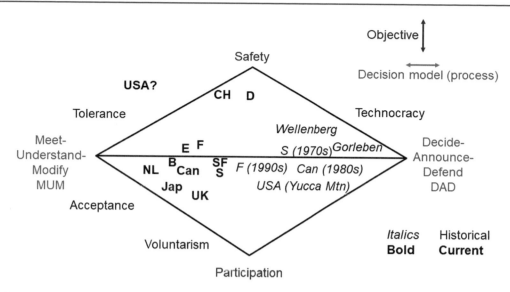

Fig. 5.5 Shift of approaches over time in various countries: emphasis towards objectives "Safety" or "Participation" along with the antagonists of decision modelling "DAD" versus "MUM". Note that indications towards "Participation" do not deny "Safety" (using information from Blowers et al. 1991: DAD, Flüeler 2003c/Clarke 2003: MUM; Flüeler 2016; NEA, website: country profiles; own appraisal and graph) B Belgium, Can Canada, CH/Wellenberg Switzerland, D/Gorleben Germany, E Spain, F France, Jap Japan, NL Netherlands, S Sweden, SF Finland, UK United Kingdom, USA/Yucca Mtn United States

in a systematic, traceable manner. Social alignments, interests and cultural values, as required for solutions to be truly socially robust, may so be identified.

It is apparent that the process of disposition (of all wastes) will last for decades, from site characterisation via construction, emplacement, surveillance, closure, monitoring and sealing. Such a situation helps to ensure that the principles of the "chain of obligation" and of the "rolling present" are not violated: The needs of the living and secondary generations are met; the secondary and third generations will take crucial decisions with respect to project management (Figs. 3.2 and 5.6). Such a process contributes to the interest of the issue. It increases awareness and **ownership of the problem** on all levels in all societal communities (engineers, scientists, politicians, regional affected parties, etc.), and it obviously has an effect on knowledge transfer, updating archives and securing financing. There is a clear need for thoughts and action in this field, for there have been produced only a few studies on the topic (e.g., seminal like Posner 1984; Isaacs 1984; Tannenbaum 1984; Berndes and Kornwachs 1996; SKB 1996; IAEA 1999; NEA 2019). It has been acknowledged that the "context has changed greatly since the 1980s, when [Records, Knowledge and Memory, RK&M] was thought to serve the sole function of deterring intrusion into a repository. Today, the goal is to preserve information to be used by future generations while maintaining technical and societal oversight of the repository for as long as practicable", as the NEA administrator in charge of the NEA RK&M Initiative stated (Pescatore 2014). But it is not

sufficient to provide markers and other artefacts for knowledge transfer in the long run.

As the entire programme stretches over decades, a surveillance institution must be installed. As shown above, the concept of "**long-term stewardship**" was proposed against the background of waste legacies in the USA. The present idea of an oversight body was somewhat embraced by the Swiss federal authorities (BFE 2008) when starting the ongoing site selection process. The six-member "Nuclear Waste Management Advisory Board" is pluralistically composed, with one representative of the nuclear industry, others are independent experts from the geological disposal community, ethics and media. It is chaired by a member of the National Council (House of Representatives). The anti-nuclear NGOs refused to collaborate but they do participate in Sectoral-plan bodies. Even though its mandate is to "offer views from an outside perspective" and to "help identify process risks and obstacles at an early stage" (BFE 2022, website), the board has not played a major role in solving conflicts so far. At any rate, and overall, however, the traditional strategy of linear decision making (by the authorities and proponents) is tentatively superseded by a dynamic and integrative, pluralistic model. Such a debate has not been launched so far with regard to CO_2. One issue is that much of the **CCS** activities and research, including site selection, characterisation and engineering designs, are proprietary and thus not disclosed to a wider public (Dean et al. 2020).

Thus, a possible generic way forward might be the following: On the initiative of such a conceptual body, with a broad composition of societal concerns, a discourse takes

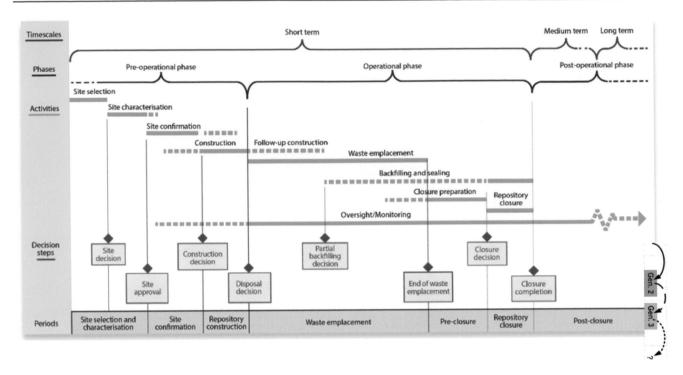

Fig. 5.6 Reference time frames and examples for important activities, periods and decisions during the implementation process of deep geological radioactive waste repositories (reproduced from NEA 2019, 21)

place on principles and alternative options based thereon. Proponents, authorities and this body reach a consensus on a conceptual framework. This agreement undergoes a wide public debate. Following an inclusive consultation, the proponent elaborates a project. The authorities commission and review expertise. Eventually, the project is revised. Finally, the concerned region (the concerned public) decides on the project. The process is characterised by dynamic mutual understanding and learning, as well as drawing up the programme and project(s), respectively. A transparent decision base, with all relevant advantages and disadvantages of the options, must be provided for and openly discussed. Modifications and changes in time and context have to be considered, i.e., decision criteria may change in the course of a project, such as the appraisal of monitoring of long-term disposition facilities. Contradictory goals with all their pros and cons have to be fully put on the table. Such a procedure corresponds to the so-called **"double-loop learning"** proposed by organisational learning theory (Argyris 1976), whose requisite, however, is institutionalised programme evaluation (Leeuw et al. 1999, 3ff).

Step 2, Decide: Passive Safety and Retrievability in a Phased and Flexible Process

With regard to problem recognition, a multitude of polls in Switzerland confirmed that the majority of respondents recognised the **nuclear waste** issue as a technological

constraint (waste exists), that a solution must be found and that a domestic site would be favoured (e.g., Dichter-Institut 1992). In line with it, nine out of ten Europeans want their member states to "have a management plan for radioactive waste which specifies fixed deadlines" (EC 2019).

Way back in the 1970s, the Swiss national electorate, by a two-third majority, voted for a Federal Decree on the then Atomic Energy Act, according to which "the permanent and safe final disposition and disposal of the … radioactive wastes" have to be "guaranteed" (Federal Decree 1979). The primary goal of radioactive waste management, passive safety in a geological environment, was confirmed in the currently valid Nuclear Energy Act of 2003. The notion of final and domestic disposal has since overwhelmingly been endorsed in numerous national and regional surveys. One may, therefore, conclude that there is a consensus in the Swiss society on geological disposal as the way to go. The national Parliament passed the revised respective act prohibiting—upon the Cantonal vetoes on Wellenberg—a final vote by an eventual host canton but stipulating that "the Department [Ministry] gives a share to the host canton as well as to the neighbouring cantons and countries in direct vicinity with regard to the preparation of the general licensing decision" (Federal Nuclear Energy Act, SNEA 2003, art. 44). The option of a—final—national referendum on the ultimately chosen site is foreseen by this law.

With regard to the technical concept, a pluralistically composed expert group was initiated in 1999. It proposed the concept of "monitored long-term geological disposal", an extension of the traditional concept of final disposal by, to a certain extent, integrating controllability/monitoring and retrievability (EKRA 2000). These aspects were adopted in legislation, although clearly as secondary to passive safety. The mentioned Sectoral plan for site selection foresees three phases and is evidently in line with the US NRC's "adaptive staging" (NRC 2002), progressively becoming common in the community (NEA 2014a). Some sort of forgiveness strategy has been installed or discussed in many countries: (possible) retrievability for 300 years in France (CNDP 2022), recourses as suggested by a German expert committee (AkEnd 2002).

Internationally, a sound of retreat from disposal philosophy can be detected (especially in the USA, Japan, the UK and partially France), where the motives are manifold. They stretch from an anti-nuclear attitude via an anticipated increase in nuclear acceptance to the strategy of resource storage on behalf of subsequent reprocessing. The UK and Japan resorted to a voluntaristic approach (within so-called "nuclear communities", Fig. 5.5). If there is no goal discussion or their results are open, every stakeholder group may put forward their motives to legitimise their particular strategy. In this sense, it is understandable that the safety experts are temporising, for whom "retrievability should never be used as an excuse to make any compromise with respect to the level of scientific and technical soundness" (Zuidema 2000).

In Switzerland, most criteria for a "socially robust" procedure were considered: arguments (domestic geologic repository), evidence (one host rock, Opalinus clay), social alignments (the majority of the actors in the Sectoral plan) and interests (procedural steps such as transparency and traceability). That (main) cultural values among such divergent actors like proponents of nuclear installations and opponents may be shared is not expected.

In applying the findings on decision processes to the scope and focus of the EU research project COWAM 2, the respective Work Package chose the following structure for a comparison of countries, consisting of five parts (Flüeler 2007a):

A. Look at what was done in the PAST;
B. Find out the CONTEXT the decision-making process is in (embedding in the national policy, framing, research strategy, legislation);
C. Identify the ACTORS with their roles and responsibilities;
D. Structure the DECISION-MAKING PROCESS (with substantive and procedural principles and goals);
E. Trace the INVOLVEMENT OF SOCIETY.

This may be visualised as follows (Fig. 5.7, overleaf):

A   THE PAST: Experience

B   THE CONTEXT
BI   Framing: Embedding in the national policy
BII   Current official research strategy
BIII   Legislation: Embedding in the regulatory system

C   THE ACTORS
CI   Formulary (institutional) stakeholders
CII   Societal stakeholders

D   DECISION MAKING
DI   Substantive principles and goals
DII   Procedural principles and rules

E   INVOLVEMENT OF SOCIETY

Many ways may lead to an expected outcome, e.g., Belgium (local partnerships), Finland (dependence on safety regulator as guardian of the process), Sweden (safety regulator and empowered municipalities), the UK (option discourse triggered by newly created body CoRWM, now Partnership with voluntary candidates), Germany (site-selection procedure recommended by a specific body, i.e., AkEnd, and newly started site selection programme), France (single-shot Public Debate exercise in 2006, focus on selected region), Switzerland (pragmatic mixture of technical approach with participation by way of spatial planning law).

Various cross-disciplinary and cross-country analyses draw similar conclusions: "Each repository not only has geological uniqueness but also has to consider cultural, historical, political and socio-economic aspects in the region" (Brunnengräber 2019, 16; cf. Fig. 6.4). This bears on virtually all dimensions: physical (the need for long-term safety of the material disposed of), technological (control versus safety/security) and social/political on multiple levels, "namely the conflict situations reflected in the landscape of complex actors, including the state, civil society and private actors" (Brunnengräber and Di Nucci 2019, 351).

The fundamental decision on **conventional hazardous waste** is being taken every time we get rid of paint, grease, car tyres, bulbs, mobiles, etc. "Special waste" is a fuzzy term and has to be handled in a special way (cf. definitions in Sect. 4.2). Some are recycled, some are upgraded, some are incinerated and some have to be deposited—about half of the hazardous waste in the European Union is landfilled (Fig. 5.8).

In the **CCS** field, engagement with the public is limited even though similar perception issues as in the nuclear field

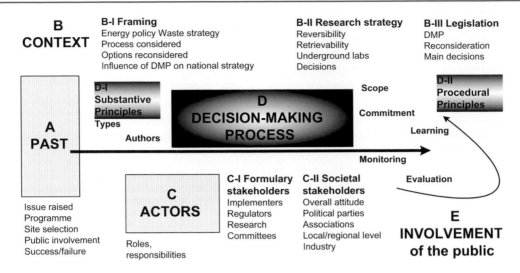

Fig. 5.7 Decision-making process criteria applied in the EU research project COWAM 2 to 12 European countries. DMP Decision-making process (reproduced from Flüeler 2007a)

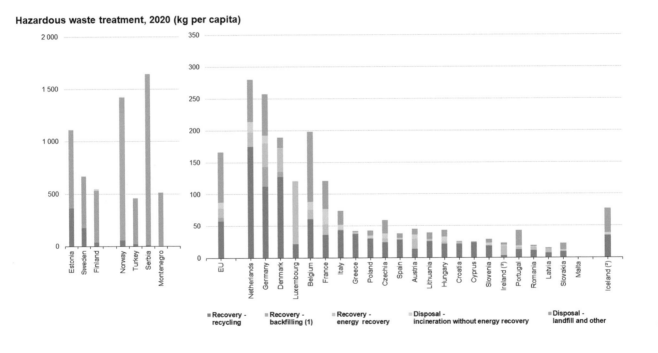

Fig. 5.8 Conventional toxic waste produced in Europe, "energy recovery" means incineration (reproduced from Eurostat 2020)

exist and have to be addressed: Not in (or under) My Back Yard, NIMBY (Krause et al. 2014), process transparency, lack of mutual trust, respect and understanding (Feenstra et al. 2010), etc. Some engineers conclude that "future leaks are as good as ruled out" once the (injection) well is closed and decommissioned and the CO_2-proof sealing has been successfully carried out, recognising that after closure monitoring possibilities are limited anyway. And "there is no reason to assume that leakage will develop after injection is completed" (Read et al. 2019, 10). The CCS Directive of the European Union allows to reduce the (minimum) leakage control period of 20 years and, consequently, transfer all responsibility to the competent authorities (EC 2009). In the case of the ROAD project, near Rotterdam, public engagement was organised by a "dedicated Stakeholder Management team" which "prevented that the project would be locked in a purely technical tunnel vision" (Read et al. 2019, 11).

Step 3, Implement: Well Started—But Prepared for the Long Term?

After more than a decade of Sectoral plan, at the start of the decisive final phase 3, one may say that the Swiss site-selection process for **radioactive waste** disposal is and continues to be long and tough (longer than initially planned by the Office of Energy), but it is ongoing; there are no major dropouts, critical comments were recognised and the work has been improved. The players recognise each other to contribute their share (Flüeler 2014d, 2015). They have to plan for the real long term, also staffwise and institutionally, and need persistence and stamina. Whether they are prepared for the programme's longevity remains to be seen: find the "right" location, build, operate, close, post-monitor the site, and all this in a positive national to local embedding.

Conventional hazardous waste is led into many different material streams and treated in many ways. As its definition is very fuzzy, i.e., sweeping, it is illustrative to look into well-itemised waste such as from electrical and electronic equipment: TV sets, computers, cell phones, ovens, refrigerators, etc. Globally, this waste field grows dramatically (https://globalewaste.org/map/), and only one-fifth is currently recycled even though the collection rate is relatively high in major producing regions (e.g., 42.5% in Europe in 2019 according to Forti et al. 2020, 76). Some of it, unfortunately, is exported and its track is lost (Fig. 5.9). Worldwide, we lose secondary raw material amounting to 57 billion USD, as many goods contain gold, silver, palladium, copper and other precious substances. A tonne of mobile phone waste alone is worth 10,000 EUR (ibid.).

The Norwegian **CCS** projects Sleipner and Snøvit in the Barents Sea have a "uniquely long track record of experience", namely 22 years (Ringrose 2018), with a predicted CO_2 dissolution of 10% within two decades. Boundary Dam (Canada) and Petra Nova (USA) are the only two commercial-scale (retrofit) coal-fired power plants currently operational in the world that use CCS technology, evidently with no sufficient database (Mantripragada et al. 2019).

5.7   (Instead of) Policy Evaluation: A Quick Glimpse at SWOTs

Technical assessments are continually undertaken in the Swiss nuclear site-selection procedure (e.g., Leuz and Rahn 2014). But full-fledged policy evaluations, whether continuous or periodic, are rare, in fact just punctual up to this point (like Planval 2014, on regional participation). To give a current impression at least, the following presentation (Flüeler 2014c, 2015) reflects on a quick-and-dirty profile of Strengths, Weaknesses, Opportunities and Threats rendered by members of major players (SWOT analysis, Kotler et al. 2010, Fig. 5.10 in table form). *Strengths*: overall, the Sectoral plan is a suitable instrument with adequate flexibility; the players stick to their given roles; safety first is held up; transparency and traceability are practised. *Weaknesses*: asymmetry of players; concerned regions without right of decision; motivation sinking; frustration rising. Chances (*Opportunities*): freedom of action. Risks (*Threats*): clash of interests; diverse levels of state, introduction of veto right for stakeholders. In detail,

statements of representatives of six main players are rendered below:

– Lead regulator (conduct of Sectoral plan): Federal Office of Energy
– Technical regulator: ENSI
– Implementer: Nagra
– Cantons (joint statement by project managers from potential siting cantons)
– Regions (joint statement by chairmen of regional conferences in potential siting regions)
– Freelance expert

Strengths (+) of Swiss Sectoral Plan

Lead regulator
Comprehensive coordination; division of roles, participation of most important players, sufficient flexibility to incorporate new evidence and requirements; knowledge transfer to siting cantons and regions.

Technical regulator
Safety-related criteria and stepwise procedure as base; causality principle as legal basis for procedure and participating institutions, clear division of roles; independent competence of the federal level, synchronisation with disposal programme.

Implementer
Clear legal basis (on disposal concept as laid down in the Swiss Nuclear Energy Act (with monitoring and retrievability, SNEA 2003), site-selection process widely supported; rules, process

a Global e-waste trade network, data as of 2012

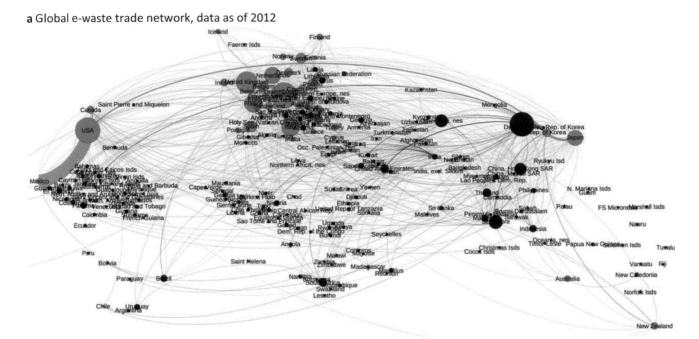

b Origins and destinations of electronic waste: the former are mainly Europe, USA, Japan, and Australia, the latter China and developing countries

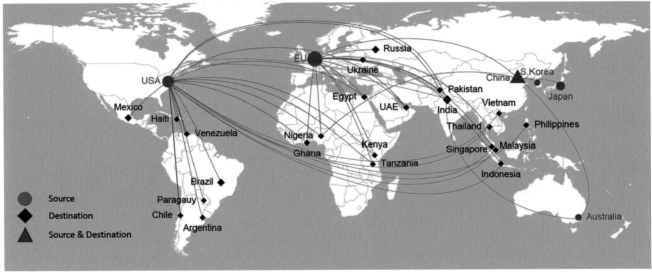

Fig. 5.9 Electronic and electric waste flows (reproduced from: **a** Lepawsky 2015, 155 (excerpt), **b** Okeme and Arrandale 2019). *Note* China has been banning imports of plastics and e-waste since

and criteria broadly discussed before Government decision, consensus on safety first, transparency; defined tasks, responsibilities and division of roles; broad and early participation of and cooperation with concerned parties (regions, cantons, neighbouring countries); intense discussions of diverse issues.

Cantons
Sectoral plan is a suitable instrument with adequate flexibility AND safety-first paradigm, compliance with the

requirement of division of roles; transparency and traceability practised.

Regions
Know-how acquisition renders regional conferences a cross-boundary player and partner (of cantons and German districts); capability of co-setting agendas to a certain extent; critics can participate; participation makes shaky basis for regions clear.

Exterior factors	Internal capabilities	
	Strength	Weakness
	• Great government Support • Coal-dominated energy structure • Huge geologic storage potential	• Poor economic feasibility • Lack of capital source • Immature technology
Opportunities • Increasing environmental pressure • Grim energy security situation • Growing international cooperation	SO Demands for energy, security, and environmental protection are urgent, relying on the government, implementing CCS activities extensive	WO CCS projects are affected by economic and technological factors, national financial support and industrial development of new technologies should be increased, achieving broader public support
Threats • Imperfect policy and laws • Low level of public acceptance	ST The government-led Is the key, targeted laws and regulations and outreach activities of CCS implementation projects need propaganda in place	WT Financial support and the acceptance by the government and public of CCS technology are required, in order to gain policy support

SO/WO Strengths/Weaknesses and Opportunities
ST/WT Strengths/Weaknesses and Threats

Fig. 5.10 SWOT analysis of CCS development in China (reproduced from Ming et al. 2014, 615)

Weaknesses (−) of the Swiss Sectoral Plan

Lead regulator
Long-lasting and complex procedure with sufficient staff resources only; scheduling (duration prolonged), possibly resulting in demotivation and loss of know-how; preparation of information for target audience difficult; high number of actors.

Technical regulator
Technical state of knowledge in Sectoral plan not in line with public information needs; time slot to integrate cantonal experts (repaired in Stage 2).

Implementer
Pioneering procedure not rulable in detail from the beginning; high number of actors, bodies and working groups hinder traceability of procedural routines; lengthy procedure may lead to problems in understanding and to fatigue in participation; regional conferences not legitimised by elections.

Cantons
Asymmetry of players (competence, resources, power); concerned and affected regions without right of decision; sporadic signs of sinking motivation and rising frustration.

Regions
Incoherence in central issues (surface facilities, socioeconomical studies); no co-decisioning regarding safety; leading institutions not open in dialogue (access issues denied, policy decisions technocratically framed; no equal footing).

Freelance Expert
Sectoral plan with substantive construction flaws (strategic roles, methods of investigation, conceptual intractability); procedural deficits in managing criticism and in controlling the project leadership; safety issues are reduced to technical expert groups despite failing expert culture (Asse, Morsleben, Germany; WIPP, USA).

Opportunities and Threats Combined, Lessons (in the Swiss Sectoral Plan)

Continuous Tension (!?) between O and T

- Actors' preparedness to dialogue
- Unruptured handing over to (technical, political, societal) generations
- Project longevity
- Specification of requirements (criteria to narrow down)
- Management of uncertainties
- Learning and failure culture
- Are concepts/models sufficient for comparability (of geological regions)?

Conclusion

- Inclusive, systematic and participatory approach needed to single out goal priorities (presumably with safety first)
- Setting up a respective process is a prerequisite to proceed in site selection

a US Electricity users (USA at night, composite visible infrared imaging radiometer suite, VIIRS, from Suomi NPP satellite in April and October 2012)

b US Electricity producers (96 operating commercial nuclear reactors at 58 nuclear power plants in 29 states, 2019)

Locations of U.S. nuclear power plants

Source: U.S. Energy Information Administration, U.S. Energy Mapping System, April 2020

Fig. 5.11 Starting points, i.e., some boundary conditions, for site selection (Photos by **a** NASA Earth Observatory/NOAA NGDC, **b** EIA 2020, **c** US CRS 2020)

c US nuclear interim storage sites and planned final repository (80 waste sites across the nation, 2020, red dot (added, tf): 1 then planned repository in Yucca Mtn, NV). SNF Spent nuclear fuel, DOE US Dept. of Energy

Fig. 5.11 (continued)

– (National) lead agency in conjunction with
– Clear division of roles among the players
– Rules of the "game"
– Criteria to judge
  and
– Regular programme and policy evaluation (concept upcoming) mandatory to control if procedure is on track

Albeit quite generic, a SWOT analysis for CCS in China gives the picture reproduced in Fig. 5.10 (Ming et al. 2014).

5.8   Conclusions and Summary

The notion of "robustness" to qualify and stabilise a system is amplified to the more adequate dynamic approach of "resilience" and "adaptiveness". Complexity and dynamics imply that many institutions have a role to play with various functions, in a communicative atmosphere of mutual respect and learning (Chap. 6). The extended robustness concept, then, is used to specify the relatively vague concept of "**Governance**". A clear regulation of parameters and requirements as well as competencies and responsibilities of the involved parties are central. Independent regulators, on par with the implementer, must accompany and supervise all relevant steps. They need respective competence, resources and staff. It is wise to adequately involve third parties—and

the larger public—at an early stage. By that, we may avoid "institutional crisis" as well as "regulatory crises" (Kasperson et al. 1992; Hutter and Lloyd-Bostock 2017, respectively).

Example

Figure 5.11 highlights different perspectives on the siting conditions of high-level radioactive waste in the USA. What can be said regarding distributional equity?

Hints

The ones to take profit of waste generation (electricity provision) and the waste-producing units (electricity-producing plants) do not coincide with the waste sites, especially not the one repository site of Yucca Mountain to which US Congress reduced all investigations in an Amendment of 1987 to the Nuclear Waste Policy Act (US Congress 1987).

Questions

1. What is the use to move from the established concept of "robustness" to the modern idea of "resilience"?
2. Time is counted in seconds, minutes, hours, years. Why isn't there a one-and-only definition?
3. We already have Government. Why do you want governance?

Answers

1. Robustness is a well-investigated and used concept to assess the stability of technical systems. We, however, deal with complex and controversial sociotechnical systems directed towards an unknown future. It is not adequate to just "add" another layer (barrier) around the existing ones. Long-term waste programmes are, in individual human terms, open-ended and moving targets. Our modes to cope with this situation must be accordingly.

2. Correct, we count time precisely, with atomic clocks. As our living generation(s) tackle and somehow precondition future generations' environments we have to understand that our and their concepts of time may not coincide. We—necessarily—discount time, i.e., "my" 60 min right now are more valuable to me than an unknown John Doe's 60 min in March 2143.

3. We do have government, even governments, all over (policemen, taxes, regulations …). They also do govern the country. But if we are confronted with multiple actors, from local to global and back, with public and private interests, in several arenas, dealing with the climate of our descendants and waste with a half-life of 30,000 years—what then?

Exercise

Apply the SWOT technique to one of the three policy areas (radioactive waste, conventional hazardous waste, carbon storage/in a particular country):

	Helpful	Harmful
Internal	Strengths …	Weaknesses …
External	Opportunities …	Threats …

Note "Internal" means attributes of the organisation/programme itself (mostly "today" and rather under control), "External" denotes attributes of the surroundings (environment) (generally in the future and out of control). "Helpful" are positive factors, "Harmful" negative ones

Refined techniques are encouraged, e.g.:

Additional Information

All weblinks accessed 27 January 2023.

Key Readings

Key References

Chhotray V, Stoker G (2009) Governance theory and practice. A cross-disciplinary approach. Palgrave Macmillan, London. https://doi.org/10.1057/9780230583344

Flüeler T (2004a) etc.: other own references in Annex

Flüeler T (2006d) What is "long term"? Definitions and implications. In: Schneider T, Schieber C, Lavelle S (eds) Long term governance for radioactive waste management. Annex of the Final Report of COWAM 2. Work Package 4, pp 53–56. Community Waste Management 2, EURATOM/FP7, FI6W-CT-508856. COWAM2-D4-12-A. https://cordis.europa.eu/project/id/508856/reporting

Flüeler T (ed) (2007a) Decision-making processes in radioactive waste governance. Appendix: synopsis of national decision-making processes (Belgium, Czech Republic, France, Germany, Hungary, Netherlands, Romania, Slovenia, Spain, Sweden, Switzerland, United Kingdom). Work Package 3 "Quality of decision-making processes". COWAM 2. Cooperative research on the governance of radioactive waste management. Feb 2007. 72 pp

Flüeler T (2014b) Extended reviewing or the role of potential siting cantons in the ongoing Swiss site selection procedure ("Sectoral Plan"). In: NEA (ed) The safety case for deep geological disposal of radioactive waste: 2013 state of the art. Symposium proceedings. Nuclear Energy Agency, Paris, 7–9 October 2013. NEA/RWM/R (2013)9. OECD, Paris, pp 405–412. https://www.oecd-nea.org/jcms/pl_19432/the-safety-case-for-deep-geological-disposal-of-radioactive-waste-2013-state-of-the-art?details=true

Flüeler T (2015) Inclusive assessment in a site-selection process—approach, experience, reflections and some lessons beyond boundaries. In: Fanghänel S (ed) Deutsche Arbeitsgemeinschaft für Endlagerforschung (DAEF). Key topics in deep geological disposal. Conference report. Cologne, 24–26 Oct 2014. Karlsruher scientific reports, vol 7696. Karlsruher Institut für Technologie, Karlsruhe, Germany, pp 53–58 (plus slides)

Flüeler T, Blowers A (2007e) Decision-making processes in radioactive waste governance. Insights and recommendations. Work Package 3 "Quality of decision-making processes". COWAM 2. Cooperative research on the governance of radioactive waste management. Feb 2007, 26 pp. https://cordis.europa.eu/docs/projects/files/508/508856/99009291-6_en.pdf

		Opportunities	Attributes 1,2,3, to criteria a,b,c		Threats	Attributes 1,2,3, to criteria a,b,c			Number of +	Number of -
			1	2 …		1	2	3 …		
Strength										
Criteria: a			++	0		-	+	0	3	1
b			0	+		++	0	0	3	0
c …										
Weaknesses										
Criteria … xyz			-	--		+	0	-	1	4
x			0	++		--	+	++	5	2
y										
z										

Minsch J, Flüeler T, Goldblatt DL, Spreng D (2012a) Lessons for problem-solving energy research in the social sciences (Chap. 14). In: Spreng D, Flüeler T, Goldblatt D, Minsch J (eds) Tackling long-term global energy problems: the contribution of social science. Environment & Policy, vol 52. Springer, Dordrecht NL, pp 273–319

Perspectives

Ascher W (1999) Resolving the hidden differences among perspectives on sustainable development. Policy Sci 32:251–377

Governance

Beck U (1994) The reinvention of politics: towards a theory of reflexive modernization. In: Beck U, Giddens A, Lash S (eds) Reflexive modernization. Polity Press, Cambridge UK, pp 1–55
Dunn J (2000) The cunning of unreason: making sense of politics. Basic Books, New York
GSDRC, Governance Social Development Research Centre, Applied Knowledge Services (2022) University of Birmingham, Birmingham UK. https://gsdrc.org/topic-guides/inclusive-institutions/concepts-and-debates/defining-institutions/
Hutter BM, Lloyd-Bostock S (2017) Regulatory crisis. Negotiating the consequences of risk, disasters and crises. Cambridge University Press, Cambridge. https://doi.org/10.1017/9781316848012
IBRD, The International Bank for Reconstruction and Development, The World Bank (2006) A decade of measuring the quality of governance. Governance matters 2006. Worldwide governance indicators. IBRD/The World Bank Washington, DC
Kasperson RE, Golding D, Tuler S (1992) Social distrust as a factor in siting hazardous facilities and communicating risks. Social Issues 48 (4):161–187. https://doi.org/10.1111/j.1540-4560.1992.tb01950.x
Kuppler S (2012) From government to governance? (Non-) effects of deliberation on decision-making structures for nuclear waste management in Germany and Switzerland. J Integr Environ Sci 9(2):103–122
Lange Ph (2017) Sustainability governance. Exploring the potential of governance modes to promote sustainable development. Sustainable development in the 21st century, vol 1. Nomos, Baden-Baden, Germany
Ostrom E (1999) Institutional rational choice. An assessment of the institutional analysis and development framework. In: Sabatier PA (ed) Theories of the policy process. Westview Press, Boulder, Colorado, pp 35–71
Voss JP, Bauknecht D, Kemp R (2006) Reflexive governance for sustainable development. Edward Elgar, Cheltenham UK
Zürn M (2018) A theory of global governance: authority, legitimacy, and contestation. Oxford University Press, Oxford

Institutions, Learning

Argyris C (1976) Single-loop and double-loop models in research on decision making. Adm Sci Q 21:363–375
Leeuw FL, Rist RC, Sonnichsen RC (eds) (1999) Can governments learn? Comparative perspectives on evaluation and organisational learning. Routledge/Taylor and Francis, London
Meyer TL (2013) Epistemic institutions and epistemic cooperation in international environmental governance. Trans Environ Law 2(1). https://ssrn.com/abstract=2225466

North DC (1990) Institutions, institutional change and economic performance. Part of political economy of institutions and decisions. Cambridge University Press, Cambridge. https://doi.org/10.1017/CBO9780511808678
Perrow C (1982) The President's commission and the normal accident. In: Sills DL, Wolf CP, Shelanski VB (eds) Accident at Three Mile Island: the human dimensions. Westview Press, Boulder, CO, pp 173–184
Perrow C (1984) Normal accidents. Living with high-risk technologies. Basic Books, New York
Wynne B (1980) Technology, risk, and participation: the social treatment of uncertainty. In: Conrad J (ed) Society, technology, and risk assessment. Academic Press, London, pp 83–107

Closure

Bijker WE, Hughes TP, Pinch T (1987) The social construction of technological systems. New directions in the sociology and history of technology. MIT Press, Cambridge, MA. https://mitpress.mit.edu/9780262517607/the-social-construction-of-technological-systems/

Process, Time

Hansson SO (1989) Dimensions of risk. Risk Anal 9(1):107–112
Kemp R (1992) The politics of radioactive waste disposal. Manchester University Press, Manchester
Kotler P, Berger R, Bickhoff N (2010) The quintessence of strategic management. Springer, Berlin
Krütli P, Stauffacher M, Pedolin D, Moser C, Scholz RW (2012) The process matters: fairness in repository siting for nuclear waste. Soc Justice Res 25(1):79–101
Moser C, Stauffacher M, Krütli P, Scholz RW (2012a) The influence of linear and cyclical temporal representations on risk perception of nuclear waste: an experimental study. J Risk Res 15(5):459–476. https://doi.org/10.1080/13669877.2011.636836
Moser C, Stauffacher M, Krütli P, Scholz RW (2012b) The crucial role of nomothetic and idiographic conceptions of time: interdisciplinary collaboration in nuclear waste management. Risk Anal 32:138–154

Resilience, Adaptiveness/Adaptability

Andresen K, Gronau N (2005) An approach to increase adaptability in ERP systems. In: Khosrow-Pour M (ed) Managing modern organizations with information technology. Proceedings of the 2005 Information Resources Management Association International Conference. San Diego, CA, 15–18 May 2005. Idea Group Publishing, San Diego, CA, pp 883–885
Buzzanell PM (2010) Resilience: talking, resisting, and imagining new normalcies into being. J Commun 60(1):1–14. https://doi.org/10.1111/j.1460-2466.2009.01469.x
Cook BJ, Emel JL, Kasperson RE (1990) Organizing and managing radioactive waste disposal as an experiment. Policy Anal Manag 9 (3):339–366
Gross Stein J (2002) The cult of efficiency. Anansi Press, Toronto
Gunderson LH, Prichard Jr L (eds) (2002) Resilience and the behavior of large-scale systems. Scope 60. Island Press, Washington, DC
Holling CS (1996) Engineering resilience versus ecological resilience. In: Schulze PC (ed) Engineering within ecological constraints. National Academy Press, Washington, DC, pp 31–43

Holland JH (2006) Studying complex adaptive systems. J Syst Sci Complexity 19:1–8

Klinke A, Renn O (2001) Precautionary principle and discursive strategies: classifying and managing risks. Risk Res 4(2):159–173

Kovalenko T, Sornette D (2013) Dynamical diagnosis and solutions for resilient natural and social systems. GRF Davos Planet@Risk 1 (1):7–33

McCarthy IP, Collard M, Johnson M (2017) Adaptive organizational resilience: an evolutionary perspective. Curr Opin Environ Sustain 28:33–40. https://doi.org/10.1016/j.cosust.2017.07.005

Pimm SL (1984) The complexity and stability of ecosystems. Nature 307:321–326

Rip A (1987) Controversies as informal technology assessment. Knowl: Creation, Diffus, Utilization 8(2):349–371

Von Bertalanffy L (1968, 1969, [10]2006) General system theory. Foundations, development, applications. George Braziller, New York

Walker B, Holling CS, Carpenter SR, Kinzig A (2004) Resilience, adaptability and transformability in social-ecological systems. Ecol Soc 9(2):5. http://www.ecologyandsociety.org/vol9/iss2/art5/

Nuclear Waste, Long-term Stewardship

Berndes S, Kornwachs K (1996) Transferring knowledge about high-level waste repositories: an ethical consideration. 7th ICHLRWM, Las Vegas. ANS, La Grange Park, IL, pp 494–498

DOE, US Department of Energy (1999) From cleanup to stewardship. A companion report to accelerate cleanup: paths to closure. Office of Environmental Management. DOE, Washington, DC, 29 pp

DOE, US Department of Energy (2001a) Developing the report to Congress on long-term stewardship. Lessons learned and recommendations for future planning. Office of Environmental Management. DOE, Washington, DC, 98 pp

DOE, US Department of Energy (2001b) Long-term stewardship study. Volume 1—report. Final study. Office of Environmental Management. DOE, Washington, DC, 207 pp

DOE, US Department of Energy (2012) Hanford long-term stewardship program plan. DOE/RL-2010-35 Revision 1. DOE Richland Operations Office, Richland, WA

EC, European Commission (2019) Special Eurobarometer 297. Attitudes towards radioactive waste. Directorate-General for Energy and Transport, Brussels. https://data.europa.eu/euodp/en/data/dataset/S681_69_1_EBS297

IAEA, International Atomic Energy Agency (1997) Joint Convention on the safety of spent fuel management and on the safety of radioactive waste management (Waste Convention). 1997-9-5. IAEA, Vienna. https://www.iaea.org/topics/nuclear-safety-conventions/joint-convention-safety-spent-fuel-management-and-safety-radioactive-waste

IAEA (1998) Technical, institutional and economic factors important for developing a multinational radioactive waste repository. TECDOC-1021. IAEA, Vienna. https://www.iaea.org/publications

IAEA (1999) Maintenance of records for radioactive waste disposal. TECDOC-1097. IAEA, Vienna

Isaacs T (1984) The institutional dimension of siting nuclear waste disposal facilities. Office of Strategic Planning and International Programs. Office of Civilian Radioactive Waste Management. US Department of Energy, Washington, DC

KASAM (1998) The state of knowledge in the nuclear waste area in 1998. SSI News. No. 2:14–16

Kuppler S, Hocke P (2019) The role of long-term planning in nuclear waste governance. J Risk Res 22(11):1343–1356. https://doi.org/10.1080/13669877.2018.1459791

LaPorte T (2004) Elements for long term institutional stewardship in a hazardous age. Views from a 'stewardee'. Session on institutional challenges for long-term stewardship of contaminated sites. 16 February 2004. Association for the Advancement of Science, Seattle, WA

Metlay D (2021) Social acceptability of geologic disposal. In: Greenspan E (ed) Encyclopedia of nuclear energy. Elsevier, Amsterdam, pp 684–697. https://doi.org/10.1016/B978-0-12-819725-7.00157-4

NEA, Nuclear Energy Agency (2014a) Stepwise approach to decision making for long-term radioactive waste management. Experience, issues and guiding principles. NEA No. 4429. OECD, Paris

NEA (2014b) Preservation of records, knowledge and memory across generations. Monitoring of geological disposal facilities—technical and societal aspects. NEA/RWM/R(2014)2. OECD, Paris, p 54ff. https://www.oecd-nea.org/rwm/rkm/

NEA (2019) Preservation of records, knowledge and memory (RK&M) across generations. Compiling a set of essential records for a radioactive waste repository. No. 7423. OECD, Paris (NEA RK&M Initiative, 2011–2018)

NAS Commission on Geosciences, Environment and Resources (2000) Long-term institutional management of US Department of Energy legacy waste sites. National Academies Press, Washington, DC

NAS Committee on Long-Term Institutional Management of DOE Legacy Waste Sites (2003) Long-term stewardship of DOE legacy waste sites—a status report. National Academies Press, Washington, DC

NCSL, National Conference of State Legislatures (2017) Closure for the seventh generation. A report from the State and Tribal Government Working Group's Long-Term Stewardship Committee. 2017 edition. NCSL, Denver, CO. https://www.ncsl.org/research/energy/closure-for-the-seventh-generation-2017.aspx

NCSL, National Conference of State Legislatures (2022, web) Site survey responses. Interactive map. https://www.ncsl.org/research/energy/closure-for-the-seventh-generation-2017.aspx

NRC, National Research Council (2002) One step at a time. The staged development of geologic repositories for high-level radioactive waste. Board on Radioactive Waste Management. The National Academies Press, Washington, DC

Pescatore C (2014) Preservation of Records, Knowledge & Memory (RK&M) across generations. An OECD/NEA initiative under the aegis of its Radioactive Waste Management Committee (presentation). OECD, Paris

Posner R (ed) (1984) Und in alle Ewigkeit ... Kommunikation über 10000 Jahre: Wie sagen wir unsern Kindern, wo der Atommüll liegt? Zeitschrift für Semiotik 6(3):195–330

Probst KN, McGovern MH (1998) Long-term stewardship and the nuclear weapons complex: the challenge ahead. Johns Hopkins University Press, Washington, DC

SKB (1996) Information, conservation and retrieval. SKB Technical Report 96–18. Swedish Nuclear Fuel and Waste Management Company SKB, Stockholm

Strohl P (1995) Notes sur l'information du public relative aux aspects institutionels de la gestion des déchets radioactifs. In: NEA, Nuclear Energy Agency (1995) Informing the public about radioactive waste management: Proceedings of an NEA International Seminar, Rauma, Finland, 13–15 June 1995. OECD, Paris, pp 125–131

Tannenbaum PH (1984) Communication across 300 generations: deterring human interference with waste deposit sites. Technical report. Atom 01 BMI/ONWI-535. Office of Nuclear Waste Isolation. Batelle Memorial Institute, Columbus, OH

Tonn BE (2001) Institutional designs for long-term stewardship of nuclear and hazardous waste sites. Technol Forecast Soc Chang 68 (3):255–273

Torfing J (2006) Governance networks and their democratic anchorage. In: Melchior J (ed) New spaces of European governance. Conference proceedings. Research group "Governance in transition", Faculty of Social Sciences, University of Vienna, pp 109–128

Zuidema P (2000) Comments on the international situation on waste management. International Symposium on Radioactive Waste Management—Sustainable Disposal or Tentative Solutions? March 30, Bern. Forum vera, Zurich

Nuclear Waste

AkEnd (2002) Selection procedure for repository sites. Recommendations of the AkEnd—Committee on a Selection Procedure for Repository Sites, AkEnd. Dec 2002. Federal Office for Radiation Protection, BfS, Salzgitter

BFE (2008) Sectoral plan for deep geological repositories. Conceptual part. BFE (Federal Office of Energy), Bern. Consultation process: 3-2006: cantonal spatial-planning offices (draft 1); 6-06: full draft to all cantonal technical offices, Germany, Austria; 6to8-06: formal consultation (national, international); 6-,11-06: 2 workshops with organisations, political parties, documented; Summer 06: focus groups for the general public, documented; 1-2007: 1st final draft; 1 to 4-07: 1st final draft of formal consultation (180 comments, incl. 22 of 26 cantons; collective 11,300 comments); 11 to 12-07: final consultation for cantonal public administrative bodies (if contradictions in "structural (spatial) plans"), majority of cantons satisfied (by procedure); 2 Apr 2008: decision of the Federal Council (Federal Government) on the Sectoral plan. https://www.bfe.admin.ch/bfe/en/home/supply/nuclear-energy/radioactive-waste/deep-geological-repositories-sectoral-plan.html

BFE (2022, web) Website https://www.bfe.admin.ch/bfe/en/home/supply/nuclear-energy/radioactive-waste/deep-geological-repositories-sectoral-plan/nuclear-waste-management-advisory-board.html (Swiss Federal Office of Energy website)

Blowers A, Lowry D, Solomon BD (1991) The international politics of nuclear waste. St. Martin's Press, New York

Brunnengräber A (2019) Making nuclear waste problems governable. Conflicts, participation and acceptability. In: Brunnengräber A, Di Nucci MR (eds) Conflicts, participation and acceptability in nuclear waste governance. An international comparison, vol 3. Springer, Wiesbaden, pp 3–19. https://www.springerprofessional.de/en/conflicts-participation-and-acceptability-in-nuclear-waste-gover/16925830

Brunnengräber A (2019) The wicked problem of long term radioactive waste governance. In: Brunnengräber A, Di Nucci MR (eds) Conflicts, participation and acceptability in nuclear waste governance. An international comparison, vol 3. Springer, Wiesbaden, pp 335–355

CNDP (2022, web) Réponses aux questions [replies to answers, 2005–2006]. Les 3 axes de recherches. Commission particulière du débat public Gestion des Déchets Radioactifs. https://cpdp.debatpublic.fr/cpdp-dechets-radioactifs/actualite/Wcf325b50992bb.html

Dichter-Institut (1992) Schweizer Stimmbürgerinnen und Stimmbürger erwarten von Politikern und Parlamentariern tatkräftiges Vorwärtsmachen bei der Entsorgung radioaktiver Abfälle [Of policy makers, Swiss voters expect an active advancement of the disposal of radioactive wastes]. Survey. Dichter-Institut, Zürich

EKRA, Expert Group on Disposal Concepts for Radioactive Waste (2000) Disposal concepts for radioactive waste. Final report. On behalf of the Swiss Federal Department for the Environment, Transport, Energy and Communication. Federal Office of Energy, Bern. https://pubdb.bfe.admin.ch/en/suche

Espejo R, Gill A (1998) The systemic roles of SKI and SSI in the Swedish nuclear waste management system. Synchro's report for project RISCOM. SKI Report 98:4/SSI-report 98–2. SKi, SSI, Stockholm

Federal Decree [Bundesbeschluss zum Atomgesetz (BB AtG) vom 6.10.1978], Swiss national popular vote in 1979, amended in 1983, prolonged until 2010

IAEA, International Atomic Energy Agency (1994) Classification of radioactive waste. A safety guide. Safety Series 111-G-1.1. IAEA, Vienna. https://www.iaea.org/publications

Leuz AK, Rahn M (2014) The regulatory perspective: role of regulatory review for preparing and performing the Swiss site selection process. In: NEA (ed) The safety case for deep geological disposal of radioactive waste: 2013 state of the art. Symposium proceedings. Nuclear Energy Agency, Paris, 7–9 October 2013. NEA/RWM/R (2013)9. OECD, Paris, pp 53–61

NWMO, Nuclear Waste Management Organization (2005) Choosing a way forward. The future management of Canada's used nuclear fuel. Final study. NWMO, Toronto. https://www.nwmo.ca/

NWMO (2022, web) Areas no longer being studied/Study areas. https://www.nwmo.ca/en/Site-selection/Study-Areas/Areas-No-Longer-Being-Studied

Planval (2014) Aufbau der regionalen Partizipation im Sachplanverfahren zur Standortsuche von geologischen Tiefenlagern. Umsetzung und Erfahrungen [Setting up the regional participation in the Sectoral plan procedure as a site-selection search for deep geological repositories]. Bundesamt für Energie, Bern

SNEA, Swiss Federal Nuclear Energy Act of 2003-3-21. Status as of 1 Jan 2022. https://www.fedlex.admin.ch/eli/cc/2004/723/en

US Congress (1987) Amendment 1987 to the Nuclear Waste Policy Act of 1982. https://www.congress.gov/bill/100th-congress/house-bill/3430

Country Profiles

https://www.oecd-nea.org/jcms/pl_33688/radioactive-waste-management-programmes-in-nea-member-countries (nuclear issues)

https://www.oecd-nea.org/jcms/tro_6814/member-countries (NEA member countries)

https://www.world-nuclear.org/information-library/country-profiles.aspx

https://www.eea.europa.eu/data-and-maps/explore-interactive-maps/country-profiles-on-resource-efficiency (conventional waste, EU countries and cooperating countries)

Hazardous Waste

Eurostat (2022, web) Waste statistics. Hazardous waste (most recent available data: for 2020). https://ec.europa.eu/eurostat/statistics-explained/index.php/Waste_statistics#Hazardous_waste_treatment

Forti V, Baldé CP, Kuehr R, Bel G (2020) The global e-waste monitor 2020. Quantities, flows and the circular economy potential. United Nations University (UNU)/United Nations Institute for Training and Research (UNITAR)—co-hosted SCYCLE programme, International Telecommunication Union (ITU) and International Solid Waste Association (ISWA), Bonn/Geneva/Rotterdam, 120 pp

CCS

Dean M, Blackford J, Conelly D, Hines R (2020) Insights and guidance for offshore CO_2 storage monitoring based on the QICS, ETI MMV, and STEMM-CCS projects. Int J Greenhouse Gas Control 100:103120

EC, European Parliament & Council (2009) Directive 2009/31/EC of the European Parliament and of the Council of 23 April 2009 on the geological storage of carbon dioxide and amending council directive 85/337/EEC, European Parliament and Council Directives 2000/60/EC, 2001/80/EC, 2004/35/EC, 2006/12/EC, 2008/1/EC and Regulation (EC) No 1013/2006. Official J Eur Union L140:114–135

Feenstra CFJ, Mikunda T, Brunsting S (2010) What happened in Barendrecht? Case study on the planned onshore carbon dioxide storage in Barendrecht, the Netherlands. Energy research Centre of the Netherlands (ECN), Petten NL, 42 pp

Krause RM, Carley SR, Warren DC, Rupp JA, Graham JD (2014) Not in (or Under) My Backyard: geographic proximity and public acceptance of carbon capture and storage facilities. Risk Anal 34 (3):529–540. https://doi.org/10.1111/risa.12119

Mantripragada HC, Zhai H, Rubin ES (2019) Boundary Dam or Petra Nova—which is a better model for CCS energy supply? Int J Greenhouse Gas Control 82:59–68

Ming Z, Shaojie O, Yingjie Z, Hui S (2014) CCS technology development in China: status, problems and countermeasures—based on SWOT analysis. Renew Sustain Energy Rev 39:604–616

Read A, Gittins C, Uilenreef J, Jonker T, Neele F, Belfroid S, Goetheer E, Wildenborg T (2019) Lessons from the ROAD project for future deployment of CCS. Int J Greenhouse Gas Control 91:102834

Ringrose PS (2018) The CCS hub in Norway: some insights from 22 years of saline aquifer storage. Energy Procedia 146:166–172. https://doi.org/10.1016/j.egypro.2018.07.021

Website

https://www.sgi-network.org/2022/ (Sustainable governance indicators)

Sources

Augustinus Aurelius (354–430) On time. Confessiones, 397–401. XI, 14. https://www.aphorismen.de/zitat/3012 (quote Sect. 5.2)

Ausländer R (1978) Mein Atem. From: Mutterland. Gedichte, p 9. Literarischer Verlag Braun, Köln (discontinued). https://www.deutschlandfunk.de/mein-atem-heisst-jetzt-102.html (quote Sect. 2)

Clarke RH (2003) The evolution of the system of radiological protection: the justification for new ICRP Recommendations. In: Andersson K (ed) VALDOR 2003. Proceedings. Stockholm, 9–13 June 2003. SCK•CEN, SKI, SSI, NKS, OECD/NEA, UK Nirex, Stockholm, pp 1–11 (in oral presentation) (Fig. 5.5)

CRS, Congressional Research Service (2020) Nuclear waste storage sites in the United States. IF 11201. Updated April 13, 2020. https://

fas.org/sgp/crs/nuke/IF11201.pdf, https://crsreports.congress.gov/ (Fig. 5.11c)

EIA, US Energy Administration Information (2020). Nuclear explained. US nuclear industry. Last updated: April 18, 2020. https://www.eia.gov/energyexplained/nuclear/us-nuclear-industry.php (Fig. 5.11b)

Eurostat (2022, web) Waste statistics. Hazardous waste (most recent available data: for 2020). https://ec.europa.eu/eurostat/statistics-explained/index.php/Waste_statistics#Hazardous_waste_treatment (Fig. 5.8)

Flüeler T (2016) On the final report of the German Commission on Nuclear Waste Disposal. Reflections by an external observer. 2nd DAEF Conference on Key Topics in Deep Geological Disposal. Cologne, 26–28 Sep 2014. (slides) (Fig. 5.5)

Fuchs H (2016) Köln aus der Luft [aerial view of Cologne]. In: Schaefer J, Reich D, Fuchs H (eds) Köln aus der Luft: Spektakuläre Bilder der Domstadt. Regionalia Verlag/Kraterleuchten, Daun (Fig. 5.4c)

GHI, Global Health Index (2019ff) GHS index. Global health security index. Building collective action and accountability, p 310. https://www.ghsindex.org/ (Table 5.1)

Lepawsky J (2015) The changing geography of global trade in electronic discards: time to rethink the e-waste problem. Geogr J 181(2):147–159, 155. https://doi.org/10.1111/geoj.12077 (Fig. 5.9a, from https://images.app.goo.gl/JiBesHoy8ftAiyWM9)

Mammutmuseum, Niederweningen. Landscape paintings. https://www.mammutmuseum.ch/ausstellung/eiszeiten-und-klimawandel/landschaftsbilder (Fig. 5.4a, b)

NASA, https://www.nasa.gov/mission_pages/NPP/news/earth-at-night.html, https://earthobservatory.nasa.gov/features/NightLights (Fig. 5.11a)

Okeme J, Arrandale VH (2019) Electronic waste recycling: occupational exposures and work-related health effects. Curr Environ Health Rep 6:256–268. https://doi.org/10.1007/s40572-019-00255-3 (Fig. 5.9b)

Oxford English Dictionary. https://www.oed.com/ (Box 5.1)

Saint-Exupéry A (1948) Citadelle [The wisdom of the sands]. Gallimard, Paris (quote, Sect. 5.2)

Smidt R (2020) Coronavirus outbreak: what you should know about the UN's response. https://unfoundation.org/blog/post/coronavirus-outbreak-what-you-should-know-about-the-uns-response/ (quote Sect. 5.4)

TA, Recherchedesk TAmedia (2020) LOCKDOWN. Wie CORONA die Schweiz zum Stillstand brachte. Schicksale, Heldinnen und ein Bundesrat im Krisenmodus [Lockdown. How the coronavirus brought Switzerland to a halt. Destinies, heroines and a Federal Council in crisis mode]. 22 Sept. 2020. Wörterseh, Lachen SZ, Switzerland (reference Sect. 5.4)

US DoD, Department of Defense, Office of the Chief Signal Officer (1945) Aerial photo of Cologne, April 1945 (Fig. 5.4d)

WHO, Tedros AG (2020a) Director-General's opening remarks at the media briefing on COVID-19, 27 April 2020. https://www.who.int/dg/speeches/detail/who-director-general-s-opening-remarks-at-the-media-briefing-on-covid-19—27-april-2020 (quote Sect. 5.4)

WHO, Tedros AG (2020b) WHO Director-General's opening remarks at the [73rd] World Health Assembly, 18 May 2020. https://www.who.int/director-general/speeches/detail/who-director-general-s-opening-remarks-at-the-world-health-assembly (quote Sect. 5.4)

Strategic Monitoring

6

Abstract

Sustainable management of all waste fields discussed here is long term, longsome and long-standing. Strategic Monitoring is devised to see that the corresponding programmes are on track and duly implemented. It is their ongoing designing, planning, reflecting and implementing in parallel, and this along with the principles of "good governance": participatory, consensus-oriented, accountable, transparent, responsive, effective and efficient, equitable and inclusive and following the rule of law.

Keywords

Problem identification • Decision, decision making • Strategy, Strategic Monitoring • Governance • Process, procedure • Participation • Capture • Path dependence • Safety culture • Radio-active/nuclear waste • Conventional toxic/hazardous waste • Carbon Capture, Utilisation and Storage, CCUS

Learning Objectives

- Know what decision making and monitoring are about, especially strategic decision making and monitoring;
- Understand the characteristics of "good governance" and respective "decision making";
- Recognise the tight links between technology and society (culture).

6.1 Decisions

Wastes are, as we saw, factual constraints[1]—they do exist and "impose" technology on other aspects. All involved parties are confronted with this precondition. Yet, unlike other controversial technical issues, "nuclear waste policy was", in Jacob's words, "not the engine that drove politics, but the product of political, economic, and social engines which drove the politics of nuclear waste" (Jacob 1990, 22). The discourse on risk, too, is factual, and political bargaining about what is conceived as "risks" is done by the actors engaged: "Risks … are to be understood as consequences of decisions on conditions of uncertainty, social action and situations of interest", as sociologists described the negotiation of interests (Nowotny and Eisikovic 1990, 13). They amplified the technology discourse in so far as they posited "that it is not 'technical progress' resulting in changes but decisions, it is the consequences of interests and certain actions, irrespective of whether the intended consequences coincide with the factual ones or not" (ibid., 10). So let us look into what decisions and the making of decisions are.

6.1.1 What is a Decision, a Good Decision?

Via decisions, possibilities to act or alternatives become either active or passive actions. In accordance with Mag (1990), "deciding" is both selecting alternatives during a mental phase and eliciting/implementing will during a phase of realisation. Here, we assume that "choice" follows "judgement", that a decision is actually taken. Principally, decision problems are informational problems; complete information on an issue would make a debate on deciding superfluous since there would be no deviation from the initial/factual state to the final/target state—there would not be any problem to solve (Fig. 6.1). But if the information is incomplete, even variable, the question is not just what to do or not to do but whether additional information should be obtained or not. Aside from the issue of the need for a decision on actions, there is the need for a decision on information. Information is purpose-oriented knowledge in a decision situation aimed at the future; it serves to reduce the decider's uncertainty on what will actually happen in the future.

---

[1] For key terms and concepts refer to Glossary in the back.

T. Flüeler, *Governance of Radioactive Waste, Special Waste and Carbon Storage*, Springer Textbooks in Earth Sciences, Geography and Environment, https://doi.org/10.1007/978-3-031-03902-7_6

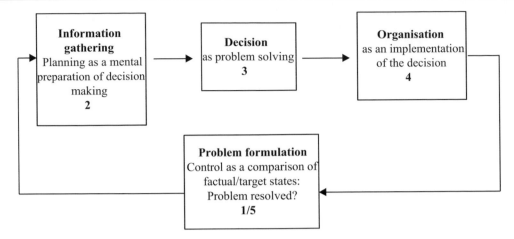

Fig. 6.1 Decisions in the feedback model of an institution (modified after Mag 1990, 3). It is important to link the iterative and cyclic decision process with the loop of Problem formulation (relation of factual/target states, **1**), Information gathering, **2**), Decision ("first" problem-solving, **3**), implementation ("Organisation", **4**) and Control (eventually "second" problem-solving, **5**) (reproduced from Flüeler 2006b, 103)

Deciding is more than the preference of an option; in decision making, one has to deal with the following questions:

- What is sufficient knowledge? Which is sufficient (for what)?
- How is sufficient knowledge collected?
- How to judge in the presence of uncertainty?
- How to integrate individual values? (discourse on dimensions, Chap. 3)
- How to assess the potential implications or side effects?
- How are the options perceived?

Thus, deciding involves the full decision-making process with problem definition, judgement, choice and implementation. So, it is not "only process" that counts, "Good" decisions are always goal-related decisions: "good" with respect to what? "Good" decisions imply good processes (which do not necessarily result in good decisions, though). A multitude of stakeholders and perspectives are involved. Therefore, particularly due to the complexity of the issues, their processes and procedures (and not only the result) are vital for the decisions to be taken. It is a stunning interaction to face.

6.1.2   What is Decision Making, Strategic Decision Making?

Decision making, therefore, is the process of deciding, the judgement made, the choice taken and, ideally, the decision implemented. With complex issues like the present ones, mostly phased collective decisions are necessary which line up in, ideally iterative, partial decisions over a long period of time. Mintzberg and colleagues, 1976, proposed a structure

of the "strategic" decision-making process as modified in Fig. 6.2 (see also Box 6.1).

Decision making is the course of action leading to a decision. It consists of several phases (Zambok and Klein 1997, also Kleindorfer et al. 1993):

- Problem identification: situation analysis (what is?), problem recognition (what is to be changed?), framing and biases, goal definition (where to?), aim (what for?);
- Options development: information gathering design (which way?), options (which preference?);
- Option selection: evaluation (on what criteria?), choice, bargaining;
- Decision;
- Implementation (setting the decision in practice);
- Evaluation (usually not included in decision making but essential for learning).

The starting point of an adequate problem-solving strategy is a thorough and thoughtful analysis of the situation concomitant with suitable system modelling. The phase of problem identification is accompanied by the formulation of goals for "optimum decisions are … always goal-oriented decisions" (Mag 1990, 28). Against this background, it is amazing that 21 out of 25 decision-making processes investigated by Mintzberg and colleagues way back in the 1970s were dominated by the problem development phase, i.e., the phase of elaborating options for solution (Mintzberg et al., 255). The authors rated it as "rather curious" that decision research was heavily focused on the evaluation choice routine (ibid., 257). Abelson and Levi detected a scarcity of research in problem recognition (Abelson and Levi 1985, 271). According to Janis and Mann, good decisions usually are characterised by careful processing of several alternatives (Janis and Mann 1977). The search

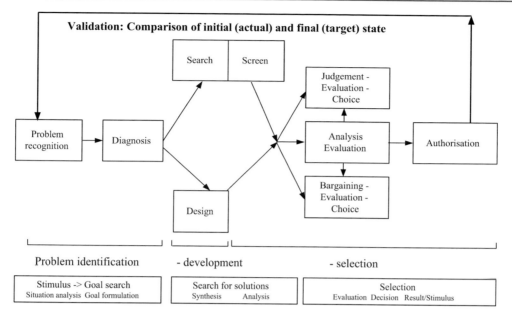

Validation: Comparison of initial (actual) and final (target) state

Fig. 6.2 General model of the strategic decision process (modified after Mintzberg et al. 1976, 266). The indications below refer to the "problem-solving cycle" in systems engineering (Haberfellner et al. 1999, 47ff). "Stimulus" denotes closing the feedback cycle, "Synthesis" means the elaboration of options, "Analysis" means the reviewing thereof. With ill-defined problems, the phases mostly are not sharply detachable (Abelson and Levi 1985, 274) (reproduced from Flüeler 2006b, 104)

routine fits the requirement of decision problems less than working out alternatives along with the design routine (Abelson and Levi 1985, 273).

Box 6.1: Strategic, Strategy

"… good strategy has an underlying structure …: 1. A diagnosis that defines or explains the nature of the challenge; 2. A guiding policy for dealing with the challenge; and 3. Coherent actions designed to carry out the guiding policy." Three important aspects of strategy include "premeditation, the anticipation of others' behavior, and the purposeful design of coordinated actions". Strategy is solving a design problem, with trade-offs among various elements that must be arranged, adjusted and coordinated, rather than a plan or choice (Rumelt 2011).

Strategy generally involves 1. setting goals and priorities, 2. determining actions to achieve the goals and 3. mobilising resources to execute the actions. A strategy describes how the ends (goals) will be achieved by the means (resources). Strategy can be intended or can emerge as a pattern of activity as the organisation adapts to its environment or competes (after Freedmann 2013; Simeone 2020).

"By and large, strategy comes into play where there is actual or potential conflict, when interests collide and forms of resolution are required. This is why a strategy is much more than a plan." (Freedman 2013, xi)

Kahneman and Tversky distinguished two decision phases: first, an editing phase where options are organised and reformulated, then, a second evaluation and choice phase (Kahneman and Tversky 1979, 1981). It is precisely in the editing phase where the evidence has to be carefully formulated, bearing a great influence on possible—divergent—appraisal: the key word here is "framing" (Tversky and Kahneman 1981). Thus, unwanted so-called context effects may be avoided. Often a decision is prepared already in the development phase when solutions are sought. The pivotal role of the process and, thus, the procedure is dealt with later on.

The following are attributes of a "good" decision-making process[2]:

- Stepwise: planning phases with milestones;
- Periodic orientation, reviewing and interim decisions: for technical and political back-up;
- Open and comprehensive option analysis;
- Iterative, with opportunities for recourse (and mutual learning);
- Reliable, accountable: unambiguous rules to be complied with (only modifiable by prior consent);
- Consistent, minimising conflicts: technical and non-technical sets of criteria;
- Coherent, continuous: for sufficient trust in "the system" (Chap. 5 and below);

---

[2] Applied to waste sites (Kunreuther et al. 1993; Easterling and Kunreuther 1995).

- Traceable: arguments and reasoning to be fully compre-hended by interested parties;
- Transparent: in broad discussion forums, aspects are put up for discussion at early stages;
- "Fair" procedure and treatment of the intra- and inter-generational equity issues, taking into account the two-fold—spatial and temporal—asymmetry: the benefit of "nuclear" electricity is broadly distributed, whereas the cost/risk of waste disposal is locally concentrated and transferred to future generations.

To reach well-supported and stable decisions, an "in-formed consent" is needed which, in turn, requires a demonstration of (all, most) possible tracks and conse-quences of actions (Fischhoff 1985).

6.1.3   What is Monitoring, Strategic Monitoring?

Deep geological repositories for nuclear waste, hazardous waste disposal sites and carbon storage/disposal objectively are long-term issues (regarding long-term safety) and require long-term institutional involvement of the technoscientific community, the waste producers, the public administration, non-governmental organisations and the general public. The demonstration of their long-term safety is avowedly very challenging and monitoring techniques may contribute to substantiate evidence, support decision making (White et al. 2017) and legitimise the respective programmes. Monitoring is the umbrella term for all types of systematical registering, measuring, observing and eventually controlling, activities and processes. What, where and when to monitor is deter-mined by its goal setting. Therefore, monitoring may be operational, confirmatory (in the near field) or environmental (in the far field) (Fig. 6.3). Strategic Monitoring, as proposed in this contribution and in addition to the above, may con-tribute to process, implementation or policy and institutional surveillance.[3]

Plans are worthless, but planning is everything.

Dwight D Eisenhower, 15 November 1957, New York Times

The "preservation of records, knowledge and memory across generations" as labelled by the corresponding Nuclear Energy Agency, NEA initiative (NEA 2019a) should encompass the tailored transfer of knowledge, concept and system understanding, insights, experience and documenta-tion to specific audiences such as above. This is laudable but focused on geologically long-term dimensions as well as static records and hardware artefacts. Yet, probably the more crucial programme phase rests before the waste is "aban-doned" in a presumably safe geologic environment, i.e., in the very decades to come (Fig. 3.2). Strategic Monitoring is devised to be an integrative and dynamic tool of targeted, yet adaptive, management until the waste can be left to itself. According to Freedman (2013), strategy is about "main-taining a balance between ends, ways and means; about identifying objectives; and about the resources and methods available for meeting such objectives", particularly "when interests collide and forms of resolution are required" (Freedman 2013, xi). One would add: as in our three cases under scrutiny. Pescatore, incidentally the former project leader of the mentioned NEA initiative, put it in a nutshell: "The current approach to legacy and, more generally, her-itage management, is [to] *create the legacy, maintain memory till the next generation, leave it to the future gen-erations to do the same*" (Pescatore 2017, emphasis origi-nal). We must hand the torch to the next generation, to the best of our knowledge and engagement, so it can get a sense of ownership. But it is their obligation to continue to pass it down to their filial generation.

6.2   Governance Applied

"Good" decisions are, as mentioned, always goal-related decisions: "good" with respect to what? Following up on Chap. 5, what is good governance? The UN assigns 8 characteristics to it (Box 6.2, also Grindle 2007).

> **Box 6.2: "Good Governance"**
> "… is participatory, consensus oriented, accountable, transparent, responsive, effective and efficient, equi-table and inclusive and follows the rule of law. It assures that corruption is minimized, the views of minorities are taken into account and that the voices of the most vulnerable in society are heard in decision making. It is also responsive to the present and future needs of society." (UNESCAP 2009, 1; see also UNDP 1997)

---

[3] Again, we have to emphasise the different time notions: 1 million years for radioactive waste and 10,000 years for CCS as isolation periods needed, several decades to around a century for implementation and surveillance in both cases.

a Operational monitoring

b Confirmatory (near-field) monitoring

c Environmental (far-field) monitoring

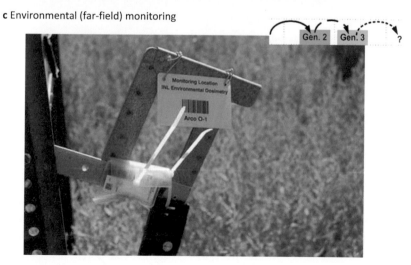

Fig. 6.3 Monitoring is designed according to object, goal and phase: **a** Operational (radiation control of spent fuel casks in interim storage), **b** Confirmatory (near field, *in situ* measurement of heating experiment in the underground lab; shown is the backfilling machine to place granular bentonite between the canister and the potential host rock Opalinus clay), **c** Environmental (far-field, above-ground measurement of toxins) (photos from: **a** Nagra 2022, **b** Mont Terri Project 2022, **c** Höser 2022)

Fig. 6.4 "Learning curve" in participation with respect to radioactive waste governance and research considering stakeholder involvement, various decision paradigms ("bubbles") and risk analysis perspectives. The shape of the curve is merely indicative (after Flüeler 2006b, 198, extended)

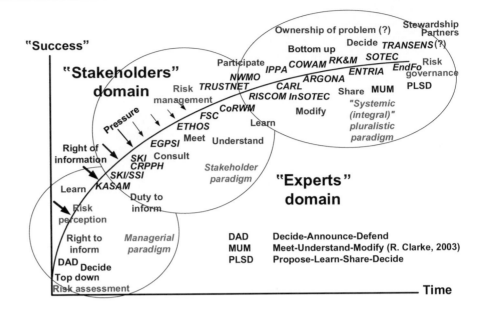

6.2.1 Collective Decision Making

Decision making paradigms evolved from the decide-announce-defend (DAD) model (Fig. 6.4, also Fig. 5.5) prevailing until the 1970s when the national radioactive waste programmes, one by one, came to a halt. In 2007, the IAEA finally acknowledged that "[m]embers of the general public and representatives of local communities recognize that they have a clear stake in the outcomes of decisions and almost always seek to have their views taken into account by the policy 'elites.' At the generic stages of the process, political acceptance seems to be the key issue. At the site-specific stages, both public and political acceptances seem to be crucial" (IAEA 2007a, 40). At any rate, we are confronted with multi-level governance issues (Hocke and Brunnengräber 2019). The following is an overview of research attempts to increasingly involve additional stakeholders (in alphabetical order):

> ARGONA (2006–2009, EU-Euratom), Arenas for risk governance: "intends to demonstrate, thereby also testing the thesis, how participation and transparency can be implemented in nuclear waste management (nwm) programmes" (ARGONA 2017).
> CARL (2006–2007), *C*itizen-stakeholders, *A*gencies responsible for radioactive waste management [RWM], social science *R*esearch organizations, and *L*icensing and regulatory authorities. Belgium, Finland, Slovenia, Sweden, UK, (Canada): "platform for interaction and collaboration and for the exchange of experiences and ideas … to study (a) decision-making processes relating to RWM; (b) how those are affected by socio-political factors; and (c) how such processes can be developed to foster stakeholder involvement while meeting regulatory requirements" (CARL 2008).
> COWAM (2000–2003, 2003), Community Waste Management (EU-Euratom), comparison of decision-making processes at the local and regional community level in nuclear waste facility siting: "will establish a common understanding of what is at stake in waste facility [siting] decision-making processes at the local and regional

community level, taking into account that the [siting] of a waste facility implies choices that are not only determined by scientific or technical options but by genuine political considerations of the relevant communities".

> COWAM 2 (2004–2006, EU-Euratom), Improving the governance of nuclear waste management and disposal in Europe: "… Better addressing and understanding societal expectations, needs and concerns as regards radioactive waste decision-making processes, notably at local and regional levels, taking advantage of the past and ongoing successful and unsuccessful experiences of RWM [Radioactive Waste Management] in the concerned European countries … Developing best practices and benchmarking on practical and sustainable decision-making processes recognised as fair and equitable by the stakeholders involved at the local, national and European levels as well as consistent on the short, medium and long term of RWM. Contributing to enable European societies to make actual progress in the governance of RWM, in order to reach practicable, accountable and sustainable decisions. COWAM 2 will aim at broad involvement of actors from civil society (with significant representation of local communities, elected representatives, and NGOs, as well as social and natural scientists from outside RWM institutions) together with the traditional actors in the field such as the implementers of RWM, the Public Authorities, experts and waste producers. COWAM 2 specifically addresses the objectives of [Euratom] Work Programme 'to better understand what influences public acceptance and develop guidance for the improved governance of geological waste disposal'" (with several work packages, see references, COWAM 2007a, b, c, Flüeler and Blowers 2007e, Hériard Dubreuil et al. 2008)
> COWAM In Practice (2007–2009, EU-Euratom), New governance approaches to radioactive waste management in Europe: Cowam in practice: "… to follow up and analyse five innovative national processes on RWM [Radioactive Waste Management: France, Romania, Slovenia, Spain, UK] on the basis of COWAM 2 results with a view to support stakeholders, particularly local communities, directly in their engagement with their particular RWM programme(s) … and to capture the learning from that experience …. to develop best practices and guidance for the application (implementation and improvement) of new inclusive governance of RWM approaches in the EU25, including benchmarking on practical and sustainable decision-making processes

recognised as fair and equitable by the stakeholders on the short, medium and long term" (CIP 2009).

CoRWM (2005–ongoing, CoRWM 2006) UK Committee on Radioactive Waste Management, Managing our radioactive waste safely: at the outset engaged in a comprehensive dialogue with experts and stakeholders.

CRPPH (1998–2020, disbanded), NEA Committee on Radiological Protection and Public Health: "Engagement with stakeholders for regulatory changes or for preparing and implementing the decision-making process in various situations, has been shown to be essential, but has not yet reached a consensus view as to how it should be understood, regulated, and implemented" (CRPPH 1995).

EGPSI (2001–2022), NEA Expert Group on the Process of Stakeholder Involvement in Radiation Protection Decision Making.

EndFo (2020–ongoing), Radioactive Waste Management as a Socio-Technical Project, Germany: "The transformation of nuclear waste management and the associated interactions are systematically investigated using concepts of technology assessment. The research group focuses on the societal project of realizing a repository for nuclear waste. It interprets this project and related policies as part of a 'socio-technical transformation' process".

ENTRIA (2019), Disposal Options for Radioactive Residues (2013–2018), Interdisciplinary Analyses and Development of Evaluation Principles, Germany: "contemplated the options for the disposal of high-level radioactive waste in Germany and has established ties between disciplines that are far apart. With their work on 'Governance between science and public protest', researchers at ITAS have significantly contributed to the success of the interdisciplinary and problem-oriented project".

ETHOS, European Commission project (1996–1998): "pilot project to initiate a new approach for the rehabilitation of living conditions in the contaminated territories of the Republic" Belarus (after the Chernobyl accident of 1986) (Lochard 2007).

FSC (2000–ongoing), Forum on Stakeholder Confidence of the Nuclear Energy Agency, NEA (NEA 2000, FSC/NEA 2015, 2022): "fosters learning about stakeholder dialogue and ways to develop shared confidence, informed consent and acceptance of radioactive waste management solutions".

InSOTEC (2011–2014, EU-Euratom): "country reports, which aim to identify the most significant socio-technical challenges related to geological disposal of radioactive waste in the different countries (Belgium, Canada, Czech Republic, Finland, France, Germany, Hungary, Netherlands, Slovenia, Spain, Sweden, Switzerland, United Kingdom, United States of America)" (InSOTEC 2014, 2022).

IPPA (2011–2013, EU-Euratom), Implementing Public Participation Approaches in radioactive waste disposal: "establishment of arenas where different stakeholders can move forward together to increase their understanding of the issues involved in radioactive waste disposal and of their respective views" (Czech Republic, Poland, Romania, Slovenia and Slovakia) (IPPA 2015).

KASAM (1988), Swedish Consultative Committee for Nuclear Waste Management: Ethical aspects on nuclear waste.

RISCOM (II, 2000–2003), Enhancing transparency and public participation in nuclear waste management: "integrates scientific, procedural, and organisational aspects for achieving trustworthy decision-making processes for public participation in nuclear waste management".

TRUSTNET (1994–1998, European Commission), new perspective on risk governance: "to propose more coherent, comprehensive and equitable approaches for evaluating, comparing and managing health and environmental risks" (TRUSTNET 2000, Hériard Dubreuil et al. 2002).

NWMO (2005–ongoing), Canadian Nuclear Waste Management Organization. Choosing a Way Forward: "NWMO has been engaged in a multi-year, community-driven process to identify a site where Canada's used nuclear fuel can be safely contained and isolated in a deep geological repository …. The site selection process emerged through a 2-year dialogue. It reflects the ideas, experience and best advice of a broad cross-section of Canadians who shared their thoughts on what an open, transparent, fair and inclusive process for making this decision would include".

Working in partnership (2018–ongoing), UK approach: "Communities will be right at the heart of the siting process for a GDF and a facility will be built where both a suitable site AND a willing community are selected" (UK Government 2020).

RK&M Records, Knowledge and Memory initiative (2011–2017), Nuclear Energy Agency, NEA: "encapsulated in how, through RK&M preservation, it may be possible to reduce the likelihood of inadvertent human intrusion and to support the capacities of future members of society to make their own informed decisions regarding a radioactive waste repository after closure" (NEA 2019a, 2020)

SOTEC (2017–2020), Methods and measures to deal with socio-technical challenges in storage and disposal of radioactive waste, Germany: "also social science expertise, e.g. on the socio-technical dynamics especially during the site selection process, can be of great importance in the phase of preparation and realization of measures for radioactive waste disposal".

SKB (-2020, 2022), Social research for the future, Swedish radioactive waste disposal implementer: "study the impact of the final repository project on the surrounding community and find out what the decision-making processes on such an important issue can look like …. Even though SKB considers that this research has helped to deepen understanding of the historical and economic aspects relating to the final disposal of nuclear waste as well as public opinion on the issue, it does not at the moment intend to fund new research programmes of the same kind".

SKI/SSI, Espejo and Gill (1998), The systemic roles of SKI and SSI in the Swedish nuclear waste management system.

TRANSENS (2019–2024), Transdisciplinary Research on the Disposal of High-level Radioactive Waste in Germany: "particularly concerned with questions of dialogue, equity, and ability to act in the site selection process launched in 2017. The challenging question of how to safely dispose radioactive waste cannot be solved by science alone. Social discourse and existing conflicts shape decision making in the site selection process for the underground repository, as do a number of technical challenges associated with the desired retrievability of the waste" (TRANSENS 2020).

(References and links under subheading *Social-science research projects and activities in the nuclear domain*)

DAD is dead and MUM is alive.

Roger H Clarke, former Chairman of ICRP, 2003
(MUM "Meet-Understand-Modify", DAD "Decide-Announce-Defend")

Pulling the strands together, criteria condensed from decision science and the comprehensive governance concepts (Chap. 5) operationalise what is termed "common ground" (Table 3.1) in the stepwise procedure (Sect. 3.4) and on the three discourse levels (Sect. 3.3):

Framework	Decision science (discourse levels)	Governance
Step 1	Inform yourself Information gathering	Integrated knowledge production
Discuss	**Problem recognition** Problem identification Problem formulation	Diagnosis
Step 2	Decide	
Decide	**Main goal consensus** Options Design Uncertainty handling Resilience/adaptability: reversibility, retrievability, control, pilot facility Conflict management **"Rules of the game"**	**Goals and priorities** Strong network *and* flexible structures **Procedural strategy** Rules, procedures: legislation, guidelines Determine actions: programme, resources
Step 3	Organise	Coherent action
Implement	Resilience: (regional) sense of ownership and care	Resources to execute action: adaptive institutions
(Step 4 Evaluate)	Control/validate Compare factual/target states	Oversight Long-term effects of measures Check interactive strategic development

If we indicatively apply the three discourse levels to reach "common ground" in selected waste programmes to the wider notion of governance against the background of legislation, technical and empirical perception studies, we may reach the following on the three discourse levels (Table 6.1):

– Problem recognition: There is consensus that nuclear and conventional wastes exist and have to be managed, a case not clear with respect to CCS (CCS to be implemented at all? With a bridging role? Even a Trojan horse? Storage and/or Usage? Section 2.3).

– Main goal consensus: Domestic solutions are favoured in nuclear and waste communities. The degree of protection and intervention is not unanimously defined (nuclear: no retrievability in the USA, 500 years in Germany, as long as a pilot facility is open in Switzerland). Not more than 50 years of oversight of Swiss conventional landfills are required by law. In an adequate goal analysis, the system performance strived for has to be examined as well as the so-called goal-means relations, i.e., the deployment of resources to reach the goals, and the participation in procedures (see "process utilities" below). In view of the sustainability goal relation, "protection versus control" and process- versus outcome-orientation, it is understood that the radioactive waste system has to be dynamic, adaptive and even experimental in its instruments, but not in its ultimate goal, i.e., the passive protection of present and future generations and environments. Central topics such as final disposal versus retrievability of waste (and reversibility of decisions) have to be put on the table. This is sensitive and explosive, but it has to be done, in a comprehensive way. Otherwise, it will come back to us or our descendants.

The goal hierarchy is protection over control (Table 3.1). The goal discussion has to be led in a broad and open manner, also because catchy but simplistic formulae (like the call for "reversibility of all decisions") have to be exposed and fundamental inconsistencies have to be dispelled. Impacts from unfounded decisions will likely be at the expense of future generations; inconsistencies are detrimental to the credibility of the entire system, and corrections made afterwards are at any rate expensive in view of the dimension of the programme if, at all, practical and efficient.

– Procedural strategy: As for non-experts, it is hard to get a clear view of the whole, it is easier to follow a straightforward procedure. In so far, technocrats must learn that laypersons may rather be process- than outcome-oriented. And that their own credibility is at stake; trusted procedures are interconnected with trusted and trustworthy players (e.g., Lehtonen 2020). Clear "rules of the game" (to start from scratch) were set in Switzerland (with a site-selection procedure called Sectoral plan 2008) and Germany (with Repository Site Selection Act 2013). In the USA, Congress singled out Yucca Mountain as the only site for a high-level waste repository by the 1987

Table 6.1 Accordance of selected country cases with criteria of "good governance" and "common ground" as defined above and in Table 3.1, respectively

Discourse level state of agreement (Table 3.1)	(Good) governance	Nuclear waste Switzerland	Nuclear waste Germany	Nuclear waste USA	Nuclear waste Canada	Nuclear waste Finland	Conventional waste Switzerland	CCS Netherlands
Sources (Table 6.2)	UNESCAP (2009)	SFOE, Sectoral plan (2008)/2011, SFOE (2020)	Rep. Act (2013), Kommission (2016), BGE (2020a, b)	NWTRB (2015), Politico (2020)	NWMO (2005), NWMO (2022) Nuclear Fuel Waste Act (2002)	Posiva (2018, 2022), Choi (2018)	FADWO (2020), Canton of Zurich (2020), SFOEN (2020)	Upham and Roberts (2011), Feenstra et al. (2010)
Step 1 Societal discourse								
Problem recognition	Rule of law Legislation	Waste exists Nuclear Energy Act 2005, phase out decided by national vote	Waste exists Repository Site Selection Act 2013 (2017)	Waste exists Nuclear Waste Policy Act 1982	Waste exists	Waste exists Government decision in 1983	Waste exists Env. Protection Law and Ordinance	"Waste" exists but to be disposed of?
(Organised) debate happening? (regions)	Participatory	National: votes in the past, future vote on site (ca. 2031) Regional (3): Sectoral plan	National: none, future decision in national parliament Regional: to be set up (90)	National: none Regional: none	Choosing a Way Forward (National dialogue: 2002–2005, following the criticism by Seaborn 1998)	National: none Regional: Municipal Council vote (positive in 2001)	National: none Regional: cantonal hazardous landfills (Zurich: 5)	National: none Regional: Barendrecht (1)
Step 2 Common ground								
Goal consensus	Consensus-oriented	Domestic site (surveys/law)	Domestic site (surveys/law)	Domestic site (surveys/law)	Domestic site (surveys)	Domestic site, ban on imports	Regional sites, some export	Offshore perceived safer
Step 3 Implementation								
Procedural strategy (instrumental and institutional goals)	Participatory	National vote Regional collaboration	Future decision in national parliament	Not decided	Adaptive Phased Management (2007)	Decision in principle 2001	Decision in cantonal parliament	None
	Transparent	Extensive publication	Publication	Publication		Publication		
	Accountable	National vote ca. 2031	National parliament	Congress (in indetermined future)	Site selection process (2010 +)	Parliament	Entry in Cantonal Spatial Directive Plan by parliament	
	Responsive	Several advisory bodies with reviews	After 50 years: Gorleben site withdrawn by proponent	Pros: Bush, Trump adm Cons: Obama, Trump statement (March 2020)		Not contested, on track	Local opposition	Cancelled
Evaluation	Validated	None	None, research programme	None		None	None	None
Trust in government (OECD 2021, %, rounded)		84	61	41 (Pew Research Center 2022: 20)	61	71	84	59

Table 6.2 Criteria (normal) and respective attributes (*Italics*) to monitor and evaluate institutions for an appraisal of governance and other theoretical concepts (**bold**) in order to develop Strategic Monitoring (after Flüeler 2014d, e, 2015, 2019)

Area	Approach/concept			
	"Good" governance	**Regulatory (and other) capture**	**Safety culture**	**Path dependence, lock-ins**
A. Formal (system) structure	Legitimation	Symmetry and asymmetry, respectively, (in)dependence	Continuous system learning	Persistence
	Legislation: goal, time frame, players, boundary conditions, etc.	*Research & development plan*	*Code of conduct, guidelines, etc.*	
	Degrees of participation by players/stakeholders	*Resources (staff, financial)*	*Feedback from staff and stakeholders*	*Research financing*
	Goal orientation, effectiveness and efficiency	*Competence(s) and experience*	*Education, permanent training; team learning*	*Review organisation*
	Degree of consensus, inclusiveness, capacity building	*Expert blocking*	*Organisational learning*	
	Rule of law			
B. Understanding of roles	Division of roles	Institutional analysis	(Senior management) commitment	Openness of decision making
	Programme tasks	*Interrelations with other players*	*Leadership*	*Comparison of options*
	Strategic planning	*Structure analysis*	*Employee involvement*	
	Responsibility			
C. Internal (organisational/personnel) structures	Transparency Accountability Equity	Mental models	Failure culture	Resistance *versus* innovation
	Justification of decisions	*Recurrent key statements*	*Openness of communication, culture*	*Mechanism of selection*
	Framework and respective guidelines	*Terms of reference, code of conduct*	*Trust*	*Components of self-reinforcement*
	Controlling: target analysis	*Performance analysis*	*Compliance analysis*	
	Responsiveness	*Agenda analysis*	*Incident reporting*	
	Quality management		*Complacency*	
	Reviewing		*Norms, values and basic assumptions*	

Readings in References: governance, capture, safety culture, path dependence and lock-ins; trust, participation; nuclear waste

Amendment to the Nuclear Waste Policy Act of 1982, which had foreseen a selection of sites in the east and the west based on technical criteria alone. Finland sticks out in so far as it is the only country to meet its own once-set timetable (NEA 2019b; Metlay 2021). **Conventional waste programmes** are left to the regional, in Switzerland: cantonal, level (e.g., Nelkin 1988). **CCS projects** depend on state and industrial initiatives but are legally not fully tied down (e.g., duration of liability, takeover of responsibility).

6.2.2   Players, Institutions

In view of the "co-production" of knowledge (by technoscience and society), the changing relationship of the actors as portrayed in Section 2.1 also has a bearing on the once passive "stakeholders", i.e., actors who have a role to play in the system, project, etc. They not even exert pressure on the "arrogant" expert establishment but—in participating—by taking over some of the responsibility (Fig. 6.5), up to some ownership of the problem (Fig. 6.4). Even a new epistemic community might

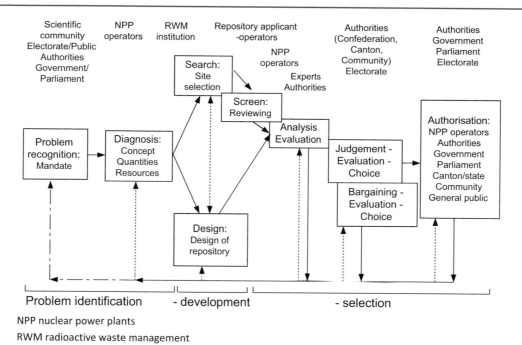

NPP nuclear power plants
RWM radioactive waste management

Fig. 6.5 Phases of the decision-making process in Swiss radioactive waste management with main stakeholders. The process of each project (interim storage, repository facilities) consists of several **stages:** from general licence to closure licence and sealing. The planned **feedbacks** (incl. criticism) are marked with dotted lines (⋯⋯). Unplanned feedbacks, such as the rejection in a referendum or financial cutbacks of waste organisations, or even new claims for concept change, are shown as broken lines (– · — · –, bottom left). At the top, the main **stakeholders** are indicated (reproduced from Flüeler 2005e)

arise when social science is included in the technical realm (Stauffacher and Moser 2010, Moser et al. 2012b).

When the DAD paradigm prevailed, one-way communication dominated, it was risk communication from the experts *to* the public. The authors of the so-called RISCOM Transparency Model (Espejo and Gill 1998) phrased it as follows: "Traditionally transparency has meant explaining technical solutions to the stakeholders and the public" (Andersson et al. 2004, 7). The official stakeholders behaved in a rather paternalistic way, according to the "deficit model", and this twofold: first with respect to communication, they wished to convey information, so to speak, *to* others (Wynne 1989, 38); second, there was some sort of "democratic deficit" (Andersson et al. 2004, 15), following the former deficit, not showing up, though, in a lack of public participation but in a lack of legitimacy of governance.

In the large variety of involved institutions, the authorities stand out. The time dimension of all three policy fields requires sufficient "institutional constancy" (Inglestam 1996). Safety-related long-term waste governance could not afford unstable institutions or vulnerable regimes as are described in studies on "failing states" (Esty et al. 1998). No other institution than a state is likely to maintain some basic stability. According to Goldstone and colleagues (2000), there are two major characteristics that affect a state's capacity to resist political crisis: the "organizational effectiveness" of that state and the legitimacy of its authority (Goldstone et al. 2000, 15).

This constellation, in turn, presupposes adequate resources, extensive reviewing, appropriate anticipatory (regulatory) research in diverse technical, non-technical and institutional fields as well as a continuous international and intergenerational knowledge transfer (see e.g., IAEA 1998).

6.2.3 Rules, Procedures, Process

In the old times of "technical fix" when nuclear issues were perceived as straightforward scientific and engineering tasks (Metlay 1978), procedures were equally clear: with license applications and permits if all regulatory requirements were met. Meanwhile, even in the nuclear community, it has been recognised that processes are much more intricate (e.g., King et al. 1991, Solomon et al. 2010, Metlay 2016, NEA 2004b, c, 2017). Clear roles and responsibilities of stakeholders are required, also a stepwise approach that is reversible and an intensive dialogue with stakeholders and the public (Pescatore and Vári 2006; NEA 2004a, 2020).

As laypeople cannot, by definition, fully understand the expert community's arguments, they have to rely on (their trust in) experts and the process. Whereas project proponents focus on technical benefits (e.g., good leakage behaviour), concerned and affected parties may prioritise "process utilities" (Sowden 1984, 297), viz., their involvement in the procedure. The so-called policy style comes in: type of problem-solving procedure (openness, transparency), type of stakeholder contact (communication behaviour, type of negotiation) and problem-solving behaviour (target consideration, problem relevance, degree of activity, planning perspective). In short, process matters. Kleindorfer and colleagues coin it as follows: "Good decisions result from [a] sound process, in much the same way that great golf or tennis shots result from great swings" (Kleindorfer et al. 1998, 392). A representative of the Physicians for Social Responsibility, PSR went even farther in 1995: "… proving the long-term safety of repositories is not a matter of the actual feasibility of a site but whether the proof submitted can be accepted by the institutions and the general public participating in the procedure" (Nidecker 1995, transl. tf).

Rip's (1987) cited definition of "social robustness" (Sects. 3.4, 5.1 and 5.6) sheds light on the circumstance that "process" is not just an issue of participation (the more "stakeholder involvement" the better), but that many aspects from diverse perspectives (and fields) are and should be expounded. The inherent emphasis on the process is a productive approach to eventually integrate different aspects and perspectives. The Sixth European Framework Programme for Research and Technological Development, launched back in 2002, stated that "[r]esearch alone cannot ensure societal acceptance; however, it is needed in order to … promote basic scientific understanding relating to safety and safety assessment methods, and to develop decision processes that are perceived as fair and equitable by the stakeholders involved" (CEC 2002, 187ff).

Procedures symbolise the continuity of similar experiences and may add to actors retaining and gaining trust in the political system. Luhmann postulated that institutions would produce their own legitimacy via procedures (Luhmann 1969, 199). Meanwhile, research has shown that these must be perceived as fair and the respective authorities must be trusted (Machura 2017; Krütli et al. 2010a, b, 2012). Long-term waste discussions are paternalistic in the sense that we (today's generations) decide over our descendants, whether we like it or not. With respect to the intragenerational asymmetry (national benefit versus regional burden), we may mitigate this ethical dilemma by appropriately including local affected and concerned stakeholders and public.

Guardian of the Process Needed

Consequently, the regulatory authorities, assisted by conceptual (advisory and oversight) bodies and ensuring continuity, have to take the lead over the *entire* programme (respecting the causality principle), including goal and programme control, concept and project management issues (e.g., in case of delays) and key economic aspects. Consequently, the upper right of the "learning curve" (Fig. 6.4) leads to the notion of stewardship and the decision model of

"Propose–Learn–Share–Decide (PLSD)": Leadership has to be exerted in conjunction with an inclusive body of knowledge, broad societal support and strong regulatory bodies, i.e., safety authorities and pluralistic oversight committees. In terms of the EU project RISCOM, the proposed "National Council for the Safe Management of Radioactive Waste" would be the "guardian of process integrity" (Andersson et al. 2004, 11, 106–108). In the author's view, it should not be the government or the parliament—as RISCOM suggested (ibid., 56)—but a pluralistically composed body, independent of the "nuclear community", yet knowledgeable about the issue and not driven by daily politics. As this body deals with comprehensive issues (from technicalities to programme evaluation and process oversight), it would be more than a steward maintaining and monitoring a facility (Kuppler and Hocke 2019; Hocke and Kuppler 2019). It would be high-level, meaning appointed by the Government, as well as accountable to Government (Fig. 7.1).

So, we have come full circle. Let us again note that our systemic analysis showed that the overall responsibility cannot lie with the one and only Panel of the Wise. And it is a matter of policy, one way or the other. The entire process with many actors, respectful and willing to learn, along with the proposed three steps (Chap. 3) will have to be both tight and flexible enough to secure a "rolling present" with an ongoing targeted yet adaptive programme.

6.2.4 Participation

More involvement and participation of stakeholders, the general public, etc., has been advocated for long. But it is an all-purpose term that needs to be specified in each application and situation. Several key questions are to be answered: Why and when should who be involved, by whom, using which technique, and, finally, what for? With which expected outcome? The public might be on board because they have the right to (normative reason), because the process and the product are better legitimised (instrumental), or because hitherto neglected issues, values and perspectives are so recognised (substantive). On top, more public participation might create and build the missing trust in institutions (e.g., Fiorino 1990, Webler et al. 1995, Beierle and Cayford 2002, Webler et al. 2006, Chilvers 2007, Reins 2017).

Even in earlier days, participation was considered useful. In concrete words, Bullard (1992) praised in the siting issue of low-level waste in the USA: "[i]ndeed, the environmental and public interest groups have been in the vanguard of public participation process, and have contributed most of the ideas overcoming the problems associated with [low-level radioactive waste] technology's complexity and reach" (Bullard 1992, 719).

Still, the benefits of public participation, in general, as well as regarding the outcome of decision-making processes, are judged controversially—and, probably, will always be. Types of participation affect the quality of the outcome; the asymmetry of power between players, legitimacy, the length and intensity of involvement leading to fatigue, etc., are never-ending issues. At any rate, before each activity, authorities, proponents, experts, stakeholders and the general public have to be aware and explicit what the "participatory means" to undertake are meant for. For example, are all truly willing to engage in mutual learning? Otherwise, this will lead to frustration on every side and, even worse, to more entrenchment than before.

Nowadays, **learning** is on everyone's lips: "Continual learning as part of a step-by-step process provides an ability to adapt the disposal process during its implementation stages such that flexibility to consider a full range of options to enhance safety, as appropriate and necessary, is considered and decisions can be reversed" (NEA 2020, 24). "Learning" is a formula used 20 times in their final report on the German site-selection procedure (Kommission 2016). The concept of continuous learning had already been statuted in the Euratom Waste Directive of 2011 (EU 2011). It has to be followed up in practice. One astounding example is the fact that Gorleben, hailed to be the site of choice in the German disposal programme for half a century, was finally eliminated by the implementer-to-be, BGE, in September 2020, as the start of the new site-selection procedure in Germany (BGE 2020b). This had been the demand by critics for years and years (Tiggemann 2019). Concerning institutional learning, the definition by Argyris (1982) is confirmed that organisational learning is a process of recognising and correcting errors (Argyris 1982). The forms of "adaptive learning" and "trial and error" dominate in the complex and politically charged radioactive waste arena.

After all, a decision is only as good as the "worst"/"weakest" partial knowledge underlying it. According to Kissling-Näf and Knoepfel (1996), "a policy … might be the more capable of learning the better-developed societal feedback and the simpler the access to novel stocks of knowledge are organised" (Kissling-Näf and Knoepfel 1996, 182, transl. tf). Structuring information and re-examining, also by special bodies, create a prolific ground to facilitate "reflexivity", "which serves to enhance the knowledge about side effects in actions by stakeholders, in politics, economy, and society" (Minsch et al. 1998, 143).

Regarding **CC(U)S** even the latest IEA flagship report does not say a word about societal concerns or wider participation. It just recognises the "critical role" of governments "to play through policies that establish a sustainable and viable market for CCUS" (IEA 2022, 16). No word about how to "identify and encourage the development of CO_2 storage in key regions" (ibid.), just "[p]ublic resistance

to storage, particularly onshore storage, has also played a role in some cases, notably in Europe" (ibid., 28). Public support is largely reduced to policy measures such as carbon tax credits, subsidies, grants, favourable regulations and state ownership (ibid., 157).

6.3 Capture, Path Dependence, Lock-Ins

As the leading regulatory authority plays a crucial role in each decision-making process (e.g., NEA 2003) they have to be particularly reflexive. Together with the operator and the affected siting region, they have to have stamina. Above all, they ought not be captured, neither by politicians nor by a technical community with vested interests. This is highly sensitive as in certain countries, e.g., Finland and possibly Sweden, trust in government and its administration is so high that people regard authorities as their "guardian of the process" (Andersson et al. 2004). An objective is to avoid "regulatory crises" (Hutter and Lloyd-Bostock 2017), but the aim should be that society may rely on their authorities. Good practice is the enforcement of a clear distinction of roles and a strong regulatory authority (e.g., IAEA 2013, 2016a).

On an even upper level, a society may be caught in technology, as most in the world are on fossil energy. A lock-in is a type of path dependency where decisions depend on prior decisions or past activities, foreclosing alternative ways (oil-fired heating keeping house owners from investing in renewables, spark-ignited cars running on petrol standing in the way of e-vehicles, etc.) (Question 2 in Chap. 2). An example on a lower level may be that a national nuclear waste disposal programme relies (persists) on a barrier concept of the 1980s (steel canisters) and invests only little resources, research, experts and money in alternatives such as ceramics.

6.4 "Soft Factors": Safety Culture, Failure Culture, Organisational Culture, Culture

We are faced with interactive complex and coupled systems that aggravate learning, processes are incompletely known, they are often unidirectionally defined and there is little room for manoeuvre. Often it is not until grave accidents provoke a change of thoughts and action. The toxic waste scandal of Love Canal in the US state of New York paved the way to change the official waste policy (Colten and Skinner 1996, Love Canal 2022). It was not before the chemical contamination of the Rhine at the Schweizerhalle incident of 1986 that a corresponding Swiss Major Accidents Ordinance was tackled (Güttinger and Stumm 1992). The nuclear accident of Three Mile Island (Harrisburg, PA) in 1979 forced the "human factor" to become the new centre of attention in reactor safety

(LaPorte 1982; Perrow 1982, 1983, 1984). And it was after the catastrophe of Chernobyl in April 1986 that the International Atomic Energy Agency, IAEA, developed the concept of **"safety culture"** (IAEA 1991, 2013, 2016b; NEA 2016). At the beginning of the same year, the most severe accident of the US space programme till then occurred, the explosion of the Challenger space shuttle 73 s after take-off. After that, NASA must have seen their Apollo 13 flight director's statement "failure is not an option" in a different light (NASA 2017, cf. Challenger 1986, Columbia disaster 2003).

Failure is always an option, sometimes even leading to improvements. Admitting failures usually means taking the blame—but it can be reassuring, and a group, a company, an institution may even emerge strengthened out of a mishap or incident if they truly learned their lessons. Examples are the shift from crystalline to sediments as host rocks in the Swiss nuclear disposal programme in the 1990s or the recent abandonment of Gorleben as a potential site in the respective German programme. In some instances, it is not clear how far learning has gone (e.g., the 2014 accidents at the US site for transuranic waste WIPP, Klaus 2019, Ewing et al. 2016). Recognising the longevity of the processes, each information and knowledge transfer process may produce errors, misunderstandings and misinterpretations.

Following the concept of this contribution (integrate issues and perspectives), the notion of **"failure culture"** applies on several levels:

- *Conceptually:* implement a robust site selection, allow regress if considered essential;
- *Regulatory:* execute safety assessments according to phase (site selection, design, construction, operation, closure, post-closure);
- *Design-wise:* e.g., integrate control mechanisms (pilot facility for surveillance and control), ensure (limited) retrievability;
- *Organisationally, culturally:* secure information and knowledge transfer (with possible information losses as failure), assure and document ways to treat minority views, foresee enlarged assessment, install process guardian.

Safety culture, and **organisational culture** at that, encompasses the full range of levels (Box 6.3): from top management to the individual collaborator (Schein 1992; Weick 1976, 1987). According to the IAEA, it is "the assembly of characteristics and attitudes in organizations and individuals which establishes that, as an overriding priority, protection and safety issues receive the attention warranted by their significance" (IAEA 2007a, b after IAEA 1991).

> Error and the failure of many experiments are also part of the building materials for a scientific fact.
>
> Ludwik Fleck 1935

> **Box 6.3: Culture**
> "Culture refers to the cumulative deposit of knowledge, experience, beliefs, values, attitudes, meanings, hierarchies, religion, notions of time, roles, spatial relations, concepts of the universe, and material objects and possessions acquired by a group of people in the course of generations through individual and group striving." (Samovar and Porter 1994, 11)
>
> "Simply stated, culture is the rules for living and functioning in society. In other words, culture provides the rules for playing the game of life." (Samovar et al. 2012, 11)
>
> "Culture is learned … transmitted intergenerationally … symbolic … dynamic … ethnocentric." (Samovar et al. 2012, 12–13)
>
> "Culture represents our link to the past and, through future generations, hope for the future. The critical factor in this equation is communication." (Samovar et al. 2012, 12)
>
> "Culture is our common human endeavor; a historical process carried on from generation to generation that bonds us together in the community of humankind." (Romanowski 2022, web)
>
> "A society's culture consists of whatever it is one has to know or believe in order to operate in a manner acceptable to its members." (Goodenough 1957)

6.5 Conclusions and Summary

What is Strategic Monitoring? If it is not a checklist (as maintained in the Preface), what then? An evaluation of the second order? The author's vision is that it is an ongoing designing, planning, reflecting and implementing the respective (in our case: waste) programme in parallel. Done by whom? The cornerstones, in his worldview, must be societal. To me, it has long been true that the problem of sustainable management of radioactive waste is eminently driven by technology but has to be solved by society (Flüeler 2004a, 797). It cannot be done by one entity, be that called a "steward" or "guardian" or otherwise. Sustainable management of all waste fields discussed here is long term, longsome and long-standing. As no single concept or approach in one or two disciplines covers all aspects, the proposition is to set up a multifarious oversight mechanism, not meant mechanistically, which the author calls "Strategic Monitoring"—"Monitoring" is familiar to risk analysts and engineers (Fig. 6.3), while "Strategic" is a keyword for managers and policy responsibles (Box 6.1). An attempt to collocate the various aspects is given in Table 6.2. Even with all the issues dealt with in this publication, the proposal still is conceptual, and on purpose at that. Strategic Monitoring must be adaptive, open, flexible, developable and hopefully developed subsequently by readers and users. When trying to connect technical approaches with societal and political sensitivities, it not only "reveals the limits to efforts to integrate technical and social dimensions of geological disposal systems into a single formalism" (as Diaz and Ewing 2018 admit in their contribution) but also acknowledges them.

Questions

1. Is it appropriate to say or demand that Strategic Monitoring is the gold standard for the waste programmes in question?
2. What is your opinion of Samowar and Porter's dictum "Culture refers to the cumulative deposit of ..."? (Box 6.3)
3. How would you explain the notion "deficit model"?

Answers

1. To adopt such an attitude would be missing the point. "Gold standards" and the like are focused assessment vehicles for defined projects. Evaluations (ex-ante, ongoing, ex-post) are necessary but not sufficient because Strategic Monitoring deals with all relevant aspects of the respective decades-long programme (from discussing the main pillars to deciding on each milestone).
2. In the author's view, this is a rather static formulation which falls short of what "culture" truly is. As if it were something external, just like waste to dump and walk away from. Interestingly enough, this statement is not reproduced in later editions of the reader.
3. Traditionally, experts as senders shared their knowledge with lay people as receivers (mutual learning was virtually inexistent). In parallel and in contradiction, this situation exhibits a lack of (democratic) legitimacy —as those lay persons presumably were the affected risk bearers (of a facility with a certain hazard potential) who should have had a say.

Additional Information

All weblinks accessed 27 January 2023.

Key Readings

Key References

Flüeler T (2004a) Long-term radioactive waste management: challenges and approaches to regulatory decision making. In: Spitzer C, Schmocker U, Dang VN (eds) Probabilistic safety assessment and management 2004. PSAM 7—ESREL '04. Berlin, June 14–18, vol 5. Springer, London, pp 2591–2596, 2593. https://link.springer.com/chapter/10.1007%2F978-0-85729-410-4_415

Flüeler T (2005b) etc.: other own references in Annex

Flüeler T, Scholz RW (2004) Socio-technical knowledge for robust decision making in radioactive waste governance. Risk, Decision and Policy 9(2):129–159. https://doi.org/10.1080/14664530490464806

Flüeler T (2001) Options in radioactive waste management revisited: a framework for robust decision making. Risk Anal 21(4):787–799. https://doi.org/10.1111/0272-4332.214150

Decisions, Strategies

Abelson RP, Levi A (1985) Decision making and decision theory. In: Lindsey G, Aronson E (eds) Handbook of social psychology, vol 1. Theory and method. Lawrence Erlbaum, New York, pp 231–309

Fischhoff B (1985) Cognitive and institutional barriers to "informed consent". In: Gibson M (ed) To breathe freely. Risk, consent, and air. Rowman & Allanheld Publishers, Totowa, NJ, pp 169–232

Freedmann L (2013) Strategy. A history. Oxford University Press, Oxford

Haberfellner R, Nagel P, Becker M, Büchel A, Von Massow H (1976, [10]1999) Systems engineering. Methodik und Praxis. Verlag Industrielle Organisation/Orell Füssli, Zürich

Janis IL, Mann L (1977) Decision making: a psychological analysis of conflict, choice, and commitment. Free Press, New York

Kahneman D, Tversky A (1979) Prospect theory: an analysis of decision under risk. Econometrica 47:263–291

Kleindorfer PR, Kunreuther HC, Schoemaker PJH (1993, [3]1998) Decision sciences. An integrative perspective. Cambridge University Press, Cambridge, UK

Mag W (1990) Grundzüge der Entscheidungstheorie [Principles of decision theory]. Franz Vahlen, München

Mintzberg H, Raisinghani D, Théorêt A (1976) The structure of "unstructured" decision processes. Adm Sci Q 21:246–275

Rumelt R (2011) Good strategy. Bad strategy. The difference and why it matters. Crown Business/Random House, New York. https://archive.org/details/goodstrategybads00rume/page/n7/mode/2up

Simeone L (2020) Characterizing strategic design processes in relation to definitions of strategy from military, business and management studies. Des J 23(4):515–534. https://doi.org/10.1080/14606925.2020.1758472

Tversky A, Kahneman D, (1981) The framing of decisions and the psychology of choice. Science (211)(Jan 30):453–458

Zambok CE, Klein G (1997) Naturalistic decision making. Lawrence Erlbaum, Mahwah, NJ

Consensus, Conflicts, Etc.

ENTRIA (2014) Memorandum zur Entsorgung hochradioaktiver Reststoffe. In: Röhlig K-J, et al (eds) Nieders. Techn. Hochschule, Hannover

Herzig, EB, Statham, ER (1993) When rationality and good science are not enough: science, politics and the policy process. In: Herzig EB, Mushkatel AH (eds) Problems and prospects for nuclear waste disposal policy. Greenwood, Westport, CT, p 10

Nowotny H, Eisikovic R (1990) Entstehung, Wahrnehmung und Umgang mit Risiken [Generation, perception, and management of risks]. Forschungspolitische Früherkennung (FER) B/34. Schweizerischer Wissenschaftsrat [Swiss Science Council] SWR, Bern

Rip A (1987) Controversies as informal technology assessment. Knowl: Creation, Diffus, Utilization 8(2):349–371

Process

Easterling D, Kunreuther H (1995) The dilemma of siting a high-level nuclear waste repository. Studies in Risk and Uncertainty, vol 5. Kluwer, Boston

Espejo R, Gill A (1998) The systemic roles of SKI and SSI in the Swedish nuclear waste management system. Synchro's report for project RISCOM. SKI Report 98:4/SSI-report 98–2. SKi, SSI, Stockholm

Krütli P, Stauffacher M, Flüeler T, Scholz RW (2010b) Functional-dynamic public participation in technological decision making: site selection processes of nuclear waste repositories. J Risk Res 13(7):861–875. https://doi.org/10.1080/13669871003703252

Krütli P, Stauffacher M, Pedolin D, Moser C, Scholz RW (2012) The process matters: fairness in repository siting for nuclear waste. Soc Justice Res 25(1):79–101

Kunreuther H, Fitzgerald K, Aarts TD (1993) Siting noxious facilities: a test of the facility siting credo. Risk Anal 13(3):301–318

Luhmann N ([1969], [2]1997) Legitimation durch Verfahren. Suhrkamp, Frankfurt a. M

Machura S (2017) Legitimation durch Verfahren – was bleibt? Soziale Systeme 22(1–2):331–354

Metlay D (1978) History and interpretation of radioactive waste management in the United States. In: Bishop WP, Hoos IR, Hilberry N, Metlay DS, Watson RA (eds) Essays on issues relevant to the regulation of radioactive waste management. PB-281347. NUREG-0412. Sandia Labs, Albuquerque, NM. United States Nuclear Regulatory Commission, NRC, Washington, DC, pp 1–19

Sowden L (1984) The inadequacy of Bayesian decision theory. Philos Stud 45:293–313

Wynne B (1989) Sheepfarming after Chernobyl. A case study in communicating scientific information. Environment 31(2):10–15, 33–39

Governance

Esty DC et al (1998) The state failure project: early warning research for US foreign policy planning. In: Davies JL, Gurr TR (eds) Preventive measures: building risk assessment and crisis early warning systems. Rowman and Littlefield, Boulder, CO

Goldstone JA et al (2000) State Failure Task Force report: phase III findings. Sept. 2000. Science Applications International Corporation (SAIC). McLean, VA

Grindle ML (2007) Good enough governance revisited. Dev Policy Rev 25(5):553–574

Inglestam L (ed) (1996) Complex technical systems. Swedish Council for Planning and Coordination of Research. Affärs Litteratur, Stockholm

UNDP, United Nations Development Program (1997) Governance and sustainable human development. UNDP, New York

UNESCAP, United Nations Economic and Social Commission for Asia and the Pacific (2009) What is good governance? UNESCAP, Bangkok, Thailand

Organisations, Safety Culture, Failure Culture, Learning

Argyris C (1982) Reasoning, learning and action. Jossey-Bass Publishers, San Francisco

Challenger space shuttle disaster (1986) https://history.nasa.gov/sts51l.html

Columbia Accident Investigation Board, CAIB (2003) STS-107 Re-entry trajectory and timeline. Report volume I. August 2003. Government printing office, Washington, DC, 248 pp. https://history.nasa.gov/columbia/reports/CAIBreportv1.pdf

Columbia space shuttle disaster (2003) https://www.nasa.gov/columbia/home/index.html

Güttinger H, Stumm W (1992) An analysis of the Rhine pollution caused by the Sandoz chemical accident, 1986. Interdisc Sci Rev 17(2):127–136

Hocke P, Kuppler S (2019) Do we need a nuclear steward? Monitoring as task for a long-term governance institution. In: MODERN 2020 Consortium (ed) Development and demonstration of monitoring strategies and technologies for geological disposal. In: 2nd International Conference on Monitoring in geological disposal of radioactive waste: strategies, technologies, decision making and public involvement. Paris, 9–11 Apr 2019. Project in the Euratom research and training programme 2014–2018 under grant agreement No 662177, pp 280–284

Hutter B, Lloyd-Bostock S (2017) Regulatory crisis: negotiating the consequences of risk, disasters and crises. Cambridge University Press, Cambridge. https://doi.org/10.1017/9781316848012

IAEA, International Atomic Energy Agency (1991) Safety culture: a report by the International Nuclear Safety Group. Safety Series No. 75-INSAG-4. IAEA, Vienna. https://www-pub.iaea.org/MTCD/publications/PDF/Pub882_web.pdf

IAEA (2007a) Safety glossary, terminology used in nuclear safety and radiation protection. IAEA, Vienna. https://www-pub.iaea.org/MTCD/publications/PDF/Pub1290_web.pdf

IAEA (2013) Regulatory oversight of safety culture in nuclear installations. TECDOC-1707. IAEA, Vienna

IAEA (2016a) Governmental, legal and regulatory framework for safety. General Safety Requirements No. GSR Part 1 (Rev. 1). IAEA, Vienna. https://www.iaea.org/resources/safety-standards

IAEA (2016b) Performing safety culture self-assessments. Safety Reports Series 83. IAEA, Vienna

Kissling-Näf I, Knoepfel P (1996) Lernfiguren in der schweizerischen Umweltpolitik [Learning types in Swiss environmental policy]. In: Roux M, Bürgin S (eds) Förderung umweltbezogener Lernprozesse in Schulen, Unternehmen und Branchen. Birkhäuser, Basel, pp 159–184

Klaus DM (2019) What really went wrong at WIPP: an insider's view of two accidents at the only US underground nuclear waste repository. Bull At Scientists 75(4):197–204. https://doi.org/10.1080/00963402.2019.1628516

LaPorte TR (1982) On the design and management of nearly error-free organizational control systems. In: Sills DL, Wolf CP, Shelanski VB (eds) Accident at Three Mile Island: the human dimensions. Westview Press, Boulder, CO, pp 185–200

Minsch J, Feindt PH, Meister HP, Schneidewind U, Schulz T (1998) Institutionelle Reformen für eine Politik der Nachhaltigkeit [Institutional reforms for a policy of sustainability]. Springer, Berlin

Moser C, Stauffacher M, Krütli P, Scholz RW (2012b) The crucial role of nomothetic and idiographic conceptions of time: interdisciplinary collaboration in nuclear waste management. Risk Anal 32:138–154

NASA. Failure is not an option (2017, web). https://www.nasa.gov/multimedia/imagegallery/image_feature_2073.html

NEA, Nuclear Energy Agency (2016) The safety culture of an effective nuclear regulatory body. No. 7247. OECD, Paris. https://www.oecd-nea.org/upload/docs/application/pdf/2019-12/7247-scrb2016.pdf

Perrow C (1982) The President's commission and the normal accident. In: Sills DL, Wolf CP, Shelanski VB (eds) Accident at Three Mile Island: the human dimensions. Westview Press, Boulder, CO, pp 173–184

Perrow C (1983) The organizational context of human factors engineering. Adm Sci Q 28:521–541

Perrow C (1984) Normal accidents, Living with high-risk technologies. Basic Books, New York

Samovar LE, Porter RE (1994) Intercultural communication: a reader, 7th edn. Wadsworth, Boston MA, p 452

Samovar LE, Porter RE, McDaniel ER (2012) Intercultural communication: a reader, 13th edn. Wadsworth, Boston MA, p 532

Schein EH (1985, 1992) Organizational culture and leadership. Jossey-Bass, San Francisco, CA

Schweizerhalle fire after chemical factory incident (1986) https://www.bafu.admin.ch/bafu/en/home/topics/major-accidents/dossiers/schweizerhalle-chemical-accident.html, https://www.heimatkunde-muttenz.ch/index.php/29-heimatkunde/natur-und-landschaft/umwelt/49-der-grossbrand-schweizerhalle-1986 (in German)

Stauffacher M, Moser C (2010) A new 'epistemic community' in nuclear waste governance? Theoretical reflections and empirical observations of some fundamental challenges. Catalan J Commun Cult Stud 2:197–211

Weick KE (1976) Educational organizations as loosely coupled systems. Adm Sci Q 21(1):1–19

Weick KE (1987) Organizational culture as a source of high reliability. Calif Manage Rev 29(2):112–127

Trust

Lehtonen M (2020) History, trust and mistrust: lessons from radioactive waste disposal megaprojects. Project TENUMECA—the techno-politics of nuclear megaproject pathologies, economic controversies and varieties of socioeconomic appraisal. Monograph

OECD (2021, web) Trust in government (indicator). Survey 2018+. https://data.oecd.org/gga/trust-in-government.htm

Pew Research Center (2022) Americans' views of government: decades of distrust, enduring support for its role. Report as of 6 June 2022. Pew Research Center, Washington, DC. https://www.pewresearch.org

Participation

Beierle TC, Cayford J (2002) Democracy in practice. Public participation in environmental decisions. Resources for the Future, Washington, DC

Chilvers J (2007) Towards analytic-deliberative forms of risk governance in the UK? Reflecting on learning in radioactive waste. J Risk Res 10(2):197–222

Fiorino DJ (1990) Citizen participation and environmental risk: a survey of institutional mechanisms. Sci Technol Human Values 15 (2):226–243

Krütli P, Flüeler T, Stauffacher M, Wiek A, Scholz RW (2010a) Technical safety versus public involvement? A case study on the unrealized project for the disposal of nuclear waste at Wellenberg (Switzerland). J Integr Environ Sci 7(3):229–244. https://doi.org/10.1080/1943815X.2010.506879

Krütli P, Stauffacher M, Flüeler T, Scholz RW (2010b) Functional-dynamic public participation in technological decision making: site selection processes of nuclear waste repositories. J Risk Res 13(7):861–875. https://doi.org/10.1080/13669871003703252

Krütli P, Stauffacher M, Pedolin D, Moser C, Scholz RW (2012) The process matters: fairness in repository siting for nuclear waste. Soc Justice Res 25(1):79–101

Reins L (2017) Regulating shale gas. The challenge of coherent environmental and energy regulation. Leuven Global Governance series. Edward Elgar, Cheltenham Glos, UK

Webler T, Tuler S (2006) Four perspectives on public participation process in environmental assessment and decision making: combined results from 10 case studies. Policy Stud J 34(4):699–722

Webler T, Kastenholz H, Renn O (1995) Public participation in impact assessment: a social learning perspective. Environ Impact Assess Rev 15:443–463

(Regulatory) Capture, Dependency, Lock-ins

Arthur B (1989) Competing technologies, increasing returns and lock-in by historical events. Econ J 99:106–131

Carpenter D, Moss DA (2014) Preventing regulatory capture. Harvard University Press, Harvard, MA

Crouch C (1993) Industrial relations and European state tradition. Oxford University Press, Oxford

Hanson JD, Yosifon DG (2003) The situation: an introduction to the situational character, critical realism, power economics, and deep capture. Univ PA Law Rev 152:129–134

Stigler GJ (1971) The theory of economic regulation. Bell J Econ Manag Sci 2(1):3–21

Vergne JP, Durand R (2010) The missing link between the theory and empirics of path dependence: conceptual clarification, testability issue, and methodological implications. J Manage Stud 47(4):336–359

Nuclear Waste

BGE, Bundesgesellschaft für Endlagerung (2020b) Zwischenbericht Teilgebiete gemäss § 13 StandAG. BGE, Peine, Germany, 444 pp

BGE (2020a) § 36 Salzstock Gorleben. Zusammenfassung existierender Studien und Ergebnisse gemäss §§ 22 bis 24 StandAG im Rahmen der Ermittlung von Teilgebieten gemäss § 13 StandAG. BGE, Peine, Germany, 47 pp

Bullard CW (1992) Low level radioactive waste. Regaining public confidence. Energy Policy. August: 712–720

Canada Nuclear Fuel Waste Act (2002) S.C. 2002, c. 23. Minister of Justice. https://laws-lois.justice.gc.ca/eng/acts/N-27.7/

CEC, Commission of the European Communities (2002) Amended proposals for Council decisions concerning the specific programmes implementing the Sixth Framework Programme of the European … Atomic Energy Community for research and training activities (2002–2006). COM(2002) 43 final. 30 Jan 2002. Brussels. https://cordis.europa.eu/programme/id/FP6-EURATOM-NUWASTE

Choi Y (2018) Trust in nuclear companies and social acceptance of a nuclear waste repository in Finland. J Environ Inf Sci 2018–1:44–55. https://www.jstage.jst.go.jp/article/ceispapersen/2018/1/2018_44/_pdf

Diaz-Maurin F, Ewing RC (2018) Mission impossible? Socio-technical integration of nuclear waste geological disposal systems. Sustainability 10(12)4390:39. https://doi.org/10.3390/su10124390

EU (2011) Council Directive 2011/70/EURATOM. Establishing a community framework for the responsible and safe management of spent fuel and radioactive waste. Official Journal of the European Union. OJL 199, 2 Aug 2011. EU, Brussels

Ewing RC, Whittleston RA, Yardley BWD (2016) Geological disposal of nuclear waste: a primer. Elements 12(4):233–237. https://doi.org/10.2113/gselements.12.4.233

Hocke P, Brunnengräber A (2019) Multi-level governance of nuclear waste disposal. Conflicts and contradictions in the German decision making system. In: Brunnengräber A, Di Nucci MR (eds) Conflicts, participation and acceptability in nuclear waste governance. An international comparison, vol 3. Springer, Wiesbaden, pp 383–401

IAEA, International Atomic Energy Agency (1998) Technical, institutional and economic factors important for developing a multinational radioactive waste repository. TECDOC-1021. IAEA, Vienna

IAEA (2007b) Factors affecting public and political acceptance for the implementation of geological disposal. TECDOC-1566. IAEA, Vienna

Jacob G (1990) Site unseen: the politics of nuclear waste repository. University of Pittsburgh Press, Pittsburgh, PA

King GP, Katz J, Munro JF (1991) Public education, public confidence, and public acceptance of radioactive waste management facilities. In: ANL, American Nuclear Society (ed) 2nd ICHLRWM, Las Vegas, vol 1. ANS, La Grange Park, IL, pp 470–476

Kommission Lagerung hoch radioaktiver Abfallstoffe [Federal Commission on High-Level Radioactive Wastes] (2016) Abschlussbericht. Verantwortung für die Zukunft. Ein faires und transparentes Verfahren für die Auswahl eines nationalen Endlagerstandortes [Final report. Responsibility of the future. A fair and transparent procedure for the selection of a national final repository site]. K-Drs. 268. Bundestag, Berlin, 683 pp

Metlay D (2016) Selecting a site for a radioactive waste repository: a historical analysis. Elements 12(4):269–274. https://doi.org/10.2113/gselements.12.4.269

Metlay D (2021) Social acceptability of geologic disposal. In: Greenspan E (ed) Encyclopedia of nuclear energy. Elsevier, Amsterdam, pp 684–697. https://doi.org/10.1016/B978-0-12-819725-7.00157-4

NEA, Nuclear Energy Agency (2000) Stakeholder confidence and radioactive waste disposal. Workshop proceedings. Paris, 28–31 Aug 2000. OECD, Paris

NEA (2003) The regulator's evolving role and image in radioactive waste management. Lessons learnt within the NEA Forum on Stakeholder Confidence. No. 4428. OECD, Paris

NEA (2004a) Stepwise approach to decision making for long-term radioactive waste management. Experience, issues and guiding principles. No. 4429. OECD, Paris

NEA (2004b) Learning and adapting to societal requirements for radioactive waste management. OECD, Paris

NEA (2004c) Stakeholder involvement techniques: short guide and annotated bibliography. OECD, Paris

NEA (2017) International conference on geological repositories 2016. Conference synthesis. Paris, 7–9 Dec. No. 7345. OECD, Paris

NEA (2019a) Preservation of records, knowledge and memory (RK&M) across generations. Final report of the RK&M Initiative. No. 7421. OECD, Paris

NEA (2019b) Country-specific safety culture forum: Finland. NEA No. 7488. WANO, STUK, NEA. OECD, Paris

NEA (2020) Management and disposal of high-level radioactive waste: global progress and solutions. NEA No. 7532 OECD, Paris. http://www.oecd-nea.org/rwm/pubs/2020/7532-dgr-geological-disposal-radioactive-waste.pdf

Nidecker A (1995) Das Wellenbergprojekt aus der Sicht der ÄrztInnen für Soziale Verantwortung (PSR/IPPNW) Schweiz [The [Swiss] Wellenberg project in the view of IPPNW]. Stans, 28 March. PSR, Basel

NWTRB, Nuclear Waste Technical Review Board (2015) Designing a process for selecting a site for a deep-mined geologic repository for high-level radioactive waste and spent nuclear fuel: detailed analysis. NWTRB, Washington, DC

Pescatore C (2017) Information and memory for future decision making—radioactive waste and beyond. The Vision document to the workshop. In: UNESCO Chair on Heritage Futures et al. (eds) Information and memory for future decision making—radioactive waste and beyond. In: Proceedings of the Stockholm workshop 21–23 May 2019. Linnaeus University, Kalmar, Sweden, pp 3–4

Pescatore C, Vári A (2006) Stepwise approach to the long-term management of radioactive waste. J Risk Res 9(1):13–40

Politico (2020) Trump's Nevada play leaves nation's nuclear waste in limbo. The president wants to win the state he narrowly lost in 2016, but he may be jumping into an energy issue. 22 Feb 2020. https://www.politico.com/news/2020/02/22/trump-nevada-nuclear-waste-yucca-mountain-116663

Posiva (2018) Nuclear waste management at Olkiluoto and Loviisa power plants: review of current status and future plans for 2019–2021. Summary. Dec 2019. YJH-2018. Posiva Oy, n.n. p 7

Posiva (2022, web) Geological final disposal. Posiva has solved final disposal in a safe manner, which is a precondition for the operation of nuclear power plants in Finland also in the future. https://www.posiva.fi/en/index/finaldisposal/geologicalfinaldisposal.html

Repository Site Selection Act, StandAG (2013) Gesetz zur Suche und Auswahl eines Standortes für ein Endlager für Wärme entwickelnde radioaktive Abfälle (Standortauswahlgesetz) as of 5 May 2017. (BGBl. I p 1074)

Seaborn B (1998) Panel report to the Government of Canada. https://www.ceaa.gc.ca/archives/pre-2003/431C8844-1/default_lang=En_n=0B83BD43-1.html

SFOE, Swiss Federal Office of Energy (2008/2011) Sectoral plan for deep geological repositories. Conceptual part. SFOE, Bern

SFOE (2022, web) Sectoral plan for deep geological repositories. Website https://www.bfe.admin.ch/bfe/en/home/supply/nuclear-energy/radioactive-waste/waste-disposal-principles/deep-geological-repositories.html

Solomon BD, Andrén M, Strandberg U (2010) Three decades of social science research on high-level nuclear waste: achievements and future challenges. Risk, Hazards Crisis Public Policy 1:13–47. https://doi.org/10.2202/1944-4079.1036

Tiggemann A (2019) The elephant in the room. The role of Gorleben and its site selection in the German nuclear waste debate. In: Brunnengräber A, Di Nucci MR (eds) Conflicts, participation and acceptability in nuclear waste governance. An international comparison, vol 3. Springer VS, Wiesbaden, pp 69–87. https://doi.org/10.1007/978-3-658-27107-7_5

Three-Mile Island, Harrisburg NY: https://www.nrc.gov/reading-rm/doc-collections/fact-sheets/3mile-isle.html, http://tmi.dickinson.edu/, http://www.tmia.com/ (watchdog group), https://www.efmr.org/ (Citizens' radiation monitoring group)

White M, Farrow J, Crawford M (2017) Deliverable D2.1: repository monitoring strategies and screening methodologies. MOD-ERN2020, Work package 2. http://www.modern2020.eu/

Social–Science Research Projects and Activities in the Nuclear Domain

Andersson K et al (2004) Transparency and public participation in radioactive waste management. RISCOM II final report. Oct./Dec. 2003. SKI Report 2004:08. https://cordis.europa.eu/project/id/FIKW-CT-2000-00045

ARGONA (2017) Analysis for risk governance. Framework programme 6 Euratom nuclear waste. https://cordis.europa.eu/project/id/36413/de

CARL, Bergmans, A, Elam M, Kos D, Polič M, Simmons P, Sundqvist G, Walls J (2008) Wanting the unwanted: effects of public and stakeholder involvement in the long-term management of radioactive waste and the siting of repository facilities. Final report CARL project. No location, p 67

CoRWM (2006) Managing our radioactive waste safely: CoRWM's recommendations to government. Committee on Radioactive Waste Management, London. https://www.gov.uk/government/collections/corwm-position-papers

COWAM (2003) COWAM network. Nuclear waste management from a local perspective. Reflections for a better governance. https://cordis.europa.eu/project/id/FIKW-CT-2000-20072/reporting

COWAM (2000–2003) Community Waste Management. EU project. https://cordis.europa.eu/project/id/FIKW-CT-2000-20072

COWAM 2 (2004–2006) Improving the governance of nuclear waste management and disposal in Europe. European Commission. Nuclear science and technology. Directorate-General for research, Euratom. https://cordis.europa.eu/project/id/508856

COWAM 2 (ed) (2007a) Roadmap for local committee construction. Better paths towards the governance of radioactive waste. Work Package 1 "Implementing local democracy and participatory assessment methods". COWAM 2. Cooperative research on the governance of radioactive waste management, 40 pp

COWAM 2 (ed) (2007b) Long term governance for radioactive waste management. Final report of COWAM 2 Work Package 4. Work Package 4 "Long-term governance of radioactive waste". COWAM 2. Cooperative research on the governance of radioactive waste management, 67 pp

COWAM 2 (ed) (2007c) National insights. WP 5 final report. Work Package 5 "National Insights". COWAM 2. Cooperative research on the governance of radioactive waste management, 156 pp

CIP, COWAM In Practice (2009) Publishable periodic activity report. Reporting period: January 2007 to December 2009. European commission community research. CIP (Contract Number: FP6/036455). https://cordis.europa.eu/project/id/36455/reporting

CRPPH (1995), NEA Committee on Radiological Protection and Public Health. Expert Group on Stakeholder Involvement and Organisations Structures, EGSIOS. https://www.oecd-nea.org/jcms/pl_20385/committee-on-radiological-protection-and-public-health-crpph, https://www.oecd-nea.org/jcms/pl_27478/expert-group-on-stakeholder-involvement-and-organisational-structures-egsios

EGPSI (2022, web) NEA Expert Group on the Process of Stakeholder Involvement in Radiation Protection Decision Making. https://www.oecd-nea.org/jcms/pl_27554/expert-group-on-the-process-of-stakeholder-involvement-in-radiation-protection-decision-making-egpsi

EndFo (2020) Radioactive waste management as a socio-technical project. https://www.itas.kit.edu/english/rg_endfo.php

ENTRIA (2019) Disposal options for radioactive residues. Interdisciplinary analyses and development of evaluation principles. https://inis.iaea.org/search/search.aspx?orig_q=RN:51000575

Espejo R, Gill A (1998) The systemic roles of SKI and SSI in the Swedish nuclear waste management system. Synchro's report for project RISCOM. SKI Report 98:4/SSI-report 98–2. SKi, SSI, Stockholm

Flüeler T (ed) (2007a) Decision-making processes in radioactive waste governance. Appendix: synopsis of national decision-making processes (Belgium, Czech Republic, France, Germany, Hungary, Netherlands, Romania, Slovenia, Spain, Sweden, Switzerland, United Kingdom). Work Package 3 "Quality of decision-making processes". COWAM 2. Cooperative research on the governance of radioactive waste management. Feb 2007, 72 pp

Flüeler T (2007b) Decision-making processes in radioactive waste governance. Recommendations. Short version. Work Package 3 "Quality of decision-making processes". EU Concerted Action project community waste management COWAM 2, 12 pp

Flüeler T, Blowers A (2007e) Decision-making processes in radioactive waste governance. Insights and recommendations. Work Package 3 "Quality of decision-making processes". COWAM 2. Cooperative research on the governance of radioactive waste management. Feb 2007, 26 pp. https://cordis.europa.eu/project/id/508856/reporting

FSC/NEA (2015) Stakeholder involvement in radioactive waste management decision making. Annotated bibliography. Forum on Stakeholder Confidence (FSC). NEA/RWM/R(2015)4. OECD, Paris

FSC/NEA (2022, web) NEA Forum on Stakeholder Confidence. https://www.oecd-nea.org/jcms/pl_26865/forum-on-stakeholder-confidence-fsc

Hériard Dubreuil G, Bengtsson G, Bourrelier GH, Foster R, Gadbois S, Kelly GN (2002) A report of TRUSTNET on risk governance—lessons learned. Risk Res 5(1):83–95

Hériard Dubreuil G, Mays C, Espejo R, Flüeler T, Schneider T, Gadbois S, Paixà A (2008) COWAM 2. Cooperative research on the governance of radioactive waste management. Final synthesis report. European Commission. Nuclear science and technology. Directorate-General for Research, Euratom. Paris: Mutadis, 64 pp. https://cordis.europa.eu/project/id/508856/reporting

InSOTEC (2014) Addressing the long-term management of high-level and long-lived nuclear wastes as a socio-technical problem: insights from InSOTEC. European Commission Community Research. Contract number: 269906. Deliverable (D 4.1). [Universiteit Antwerpen]

InSOTEC (2022, web) https://sites.google.com/a/insotec.eu/insotec/home

IPPA (2015, web) Implementing Public Participation Approaches in radioactive waste disposal. https://cordis.europa.eu/project/id/269849/reporting

IPPA (2015, web) Toolbox. http://ippa-toolbox.oeko.de/toolboxes

KASAM (1988) Ethical aspects on nuclear waste. KASAM (Consultative Committee for Nuclear Waste Management). SKN Report 29. SKN (National Board for Spent Nuclear Fuel), Stockholm

Kuppler S, Hocke P (2019) The role of long-term planning in nuclear waste governance. J Risk Res 22(11):1343–1356. https://doi.org/10.1080/13669877.2018.1459791. (SOTEC)

Lochard J (2007) Rehabilitation of living conditions in territories contaminated by the Chernobyl accident: the ETHOS project. Health Phys 93(5):522–526. https://doi.org/10.1097/01.HP.0000285091.08936.52

NEA, Nuclear Energy Agency (ed) (2003) Stakeholder participation in decision making involving radiation: exploring processes and implications. [3rd] Workshop proceedings. Villigen, Switzerland, 21–23 Oct 2003. OECD, Paris (i.a., on CRPPH)

NEA (2020) Preservation of records, knowledge and memory (RK&M) across generations. Final report of the RK&M initiative. No. 7421. OECD, Paris. https://www.oecd-nea.org/rwm/rkm/

NWMO Nuclear Waste Management Organization (2005) Choosing a way forward. The future management of Canada's used nuclear fuel. Final study. NWMO, Toronto. https://www.nwmo.ca/en/Site-selection/About-the-Process

NWMO (2023, web) Areas no longer being studied/Study areas.https://www.nwmo.ca/en/Site-selection/Study-Areas/Areas-No-Longer-Being-Studied

RISCOM (II, 2000–2003) Enhancing transparency and public participation in nuclear waste management. Framework Programme 5. https://cordis.europa.eu/project/id/FIKW-CT-2000-00045

SKB (2022) Social research for the future. https://www.skb.com/research-and-technology/social-research/

SOTEC (2017-ongoing) Methods and measures to deal with socio-technical challenges in storage and disposal of radioactive waste, Germany. https://www.itas.kit.edu/english/projects_kupp17_sotecr.php

TRANSENS (2020) Transdisciplinary research on the disposal of high-level radioactive waste in Germany. https://www.itas.kit.edu/english/projects_hock19_transens.php

TRUSTNET, Director-General for Research (2000) The TRUSTNET framework: a new perspective on risk governance. European Commission nuclear science and technology. Brussels. European Communities, Luxembourg

UK Government (2020) Working in partnership. https://www.gov.uk/guidance/communities-and-gdf

Conventional Toxic Waste

Canton of Zurich (2020) Deponien [Landfills]. https://www.zh.ch/de/umwelt-tiere/abfall-rohstoffe/abfaelle/abfallanlagen/deponien.html (landfill types C-E)

Colten CE, Skinner PN (1996) The road to Love Canal: managing industrial waste before EPA, 1st edn. University of Texas Press, Austin, TX, p 217

FADWO, Swiss Federal Ordinance on the Avoidance and the Disposal of Waste as of 2015-12-4. Status as of 2020-4-1. SR 814.600. (English version with no legal force)

Nelkin D (1988) Risk reporting and the management of industrial crises. Manage Stud 25:341–351

Love Canal (2022, web). https://www.epa.gov/history/love-canal

SFOEN, Swiss Federal Office for the Environment (2022) Landfills. https://www.bafu.admin.ch/bafu/en/home/topics/waste/info-specialists/waste-disposal-methods/deponien.html

Carbon Capture and Storage, CCS

Feenstra CFJ, Mikunda T, Brunsting S (2010) What happened in Barendrecht? Case study on the planned onshore carbon dioxide storage in Barendrecht, The Netherlands. Energy research Centre of the Netherlands (ECN), Petten NL, p 42

IEA, International Energy Agency (2022) The role of CCUS in low-carbon power systems. https://www.iea.org/reports/the-role-of-ccus-in-low-carbon-power-systems

Upham P, Roberts T (2011) Public perceptions of CCS in context: results of NearCO2 focus groups in the UK, Belgium, The Netherlands, Germany, Spain and Poland. Energy Procedia 4:6338–6344

Websites

https://www.iea.org/ (International Energy Agency: general, CCS)

https://www.oecd-nea.org/jcms/c_12892/radioactive-waste-management/ (Nuclear Energy Agency of the OECD: radioactive waste)

https://www.ipcc.ch/ (Intergovernmental Panel on Climate Change: carbon storage)

https://www.globalccsinstitute.com (Global CCS Institute: carbon capture and storage)

Sources

Clarke RH (2003) The evolution of the system of radiological protection: the justification for new ICRP Recommendations. In: Andersson K (ed) VALDOR 2003. Proceedings. Stockholm, 9–13 June 2003. SCK·CEN, SKI, SSI, NKS, OECD/NEA, UK Nirex, Stockholm, p 1–11 (in oral presentation) (quote Sect. 6.2, Fig. 6.4)

Eisenhower DD (1957) Remarks at the national defense executive reserve conference, Nov 14, 1957 (https://www.eisenhowerlibrary.gov/media/3860). 15 November 1957, New York Times, New York

Fleck L (1935, 1979) Genesis and development of a scientific fact. Trenn TJ, Merton RK (eds), transl. Bradley F, Trenn TJ. University of Chicago Press, Chicago, IL, p 98. Orig.: Fleck L (1935, 1980) Entstehung und Entwicklung einer wissenschaftlichen Tatsache. Einführung in die Lehre vom Denkstil und Denkkollektiv. Schäfer L, Schnelle T (eds). Suhrkamp Taschenbuch Wissenschaft, Frankfurt aM, 128 pp (quote Sect. 6.4)

Goodenough W (1957) Cultural anthropology and linguistics. In: Garvin PL (ed) Report of the Seventh annual round table meeting on Linguistics and language study. Monograph Series on Language and Linguistics, no 9. Georgetown University, Washington, DC, pp 167–173 (Box 6.3)

Höser I (2022, web) https://idahoeser.inl.gov/Surveillance/TLD.html (radiation measuring device) (Fig. 6.3c)

Mont Terri Underground Laboratory (2022, web) https://www.mont-terri.ch/en/homepage.html (heating experiment) (Fig. 6.3b)

Nagra (2022, web) https://www.nagra.ch/en/types-of-radioactive-waste (spent fuel casks in the Swiss central interim storage Zwilag) (Fig. 6.3a)

Romanowski W (2022, web) Culture. https://dceac.wordpress.com/what-is-culture/ (Box 6.3)

Conclusions and Outlook

7

Abstract

Deep geological repositories of nuclear waste, long-term landfills of other hazardous waste and the substantive storage of carbon dioxide to relieve the world's climate from excessive system change are intricate and contentious policy fields with an impact of decades to hundreds of thousands of years. The present work aims to explore how society and technology can set up and implement sustainable ways—in the long run—to cope with the issues. They are long-term safety issues and require long-term institutional involvement of the technoscientific community, waste producers, public administrators, non-governmental organisations, NGOs and the public. The demonstration of long-term safety is challenging and monitoring may contribute to substantiate evidence, support decision-making and legitimise the programme. What, where and when to monitor is determined by its goal setting. Strategic Monitoring as proposed contributes to the process, implementation or policy and institutional surveillance to sustain a once launched programme. It includes the tailored transfer of knowledge, concept and system understanding, experience and documentation to specific audiences mentioned above. It is an integrative tool of targeted, yet adaptive, management and may be applicable to other long-term sociotechnical fields. Based on the analysis, suggestions for research are given.

Keywords

Strategic Monitoring • Adaptive management • Long-term governance • Radioactive/Nuclear waste • Conventional toxic/Hazardous waste • Carbon Capture and Storage, CCS • Critical thinking • Competences for sustainability • Process and policy surveillance • Tools

Learning Objectives

- Understand the presented comprehensive approach of "Strategic Monitoring" of the three waste systems (nuclear, conventional toxic and carbon storage);
- Learn which "ingredients" for literacy in Strategic Monitoring are needed: engaged experts, stakeholders, publics to set up a targeted programme; appropriate structures with a lead agency and a guardian of the process; resources, including time;
- Acknowledge (and adopt) skills for sustainable governance;
- Look into the proposed tools and roads to explore developing Strategic Monitoring.

7.1 Learning to Become Literate: The Ingredients

With the COVID-19 pandemic, we have been forced to adapt,[1] to adapt to a so-called "new normality".[2] Individuals, communities, countries and structures are affected. We are confronted with various, and contradictory, management paradigms, such as prohibition and command management thinking vs. reliance on individual self-responsibility. It is about evidence-based understanding, sound science, scientific advisory bodies supporting (or not) public health

---

[1] For key terms and concepts, refer to Glossary in the back.

[2] Unlike maintained in the media during the pandemic, the caption "normality" and "return to (full) normality" was first formulated by a former director of the Swiss nuclear regulator, now ENSI, and a former director of the French Nuclear Protection Evaluation Centre CEPN, against the experience of Chernobyl, in the 1990s. Hereby, they qualified normality in a non-normative way "a situation where the basic functions of the society operate as usual and no particular perturbation affects its members" (Prêtre and Lochard 1995, 23). Incidentally, monitoring and control play a vital part when qualifying what "normal" conditions are. For a word of caution, see Asonye (2020) and Reinhart and Rogoff (2009).

administrators with a considerate and plausible protection concept, trustworthy politicians and communicators, abiding by their own rules (or not) so that people accept recommended measures and gain skills in new normalcy (mask wearing, spatial distancing, adhering to hygiene rules, vaccinations and alertness to new variants).

Should we now be prepared for managing decades- and century-long waste programmes? Facts, figures and experiences are ambiguous. **Whom and what do we need?** We know that we all have to learn—experts, politicians, public servants, stakeholders and the public. We must recognise that unless we get a notion of common ownership of the problem(s), the enterprises are bound to fail. Expertise must be enlarged, **experts** should be scientifically rigid and modest at the same time (Box 7.1). After all, they have to recognise that, in 99.9% of their living worlds, they themselves are laypersons.

The participation of a wide range of stakeholders and the public may facilitate a national discourse that deserves that name. Actors—dedicated because they can expect to be heard—engage in a winding and steep process, supported by adequate institutional **structures** (Fig. 7.1). The **actors** should be knowledgeable, at least informed, and capable of critical thinking (see below). Institutional **resources** must be secured to reach the goals consented to (Chap. 3). A **lead agency** should handle the procedures (e.g., of siting), and a technical safety inspectorate should assess the safety analyses. A **guardian** may oversee the process to overcome possible breakpoints of government weaknesses (change of government, budget restrictions, etc.). A strategic steering body, e.g., a National Council for the Safe Management of Radioactive Waste, accompanied by scientific research, must continuously address violations of rules, the correct discourse of actors and controversial topics and the dynamics of the discourse. It should be pluralistically composed, independent of the "waste community" yet knowledgeable about the issues and the process and not driven by daily politics.

Box 7.1: Experts, Public(s)

"[The planner's] would-be solutions are confounded by a still further set of dilemmas posed by the growing pluralism of the contemporary publics, whose valuations of his proposals are judged against an array of different and contradicting scales. It should be clear that the expert is also the player in a political game, seeking to promote his private vision of goodness over others."

Rittel and Webber (1973)

"Everywhere in the world, the profession is confronted with the legacy of the past .... The time when it was

possible to manage from inside the office is over. We need to go outside and interact with the population. We have to answer unexpected and difficult questions. We are facing values, concerns and emotions that were not part of our decision-aiding models. In fact, we are challenged at all levels of our expertise .... We have to listen, and adopt a more modest and learning attitude as far as societal issues are concerned if we want to effectively take part as stakeholders in the decision-making processes in the future .... This implies both a mourning process as far as our past position is concerned and some courage to overcome the fear of change."

Lochard (2002)

"... the people whose lives are affected are the true experts on questions of value regarding the risks of technology."

Otway (1987)

"Differences in perspective or focus are due to the distinct nature of the various stakeholders. On the one hand, a delegation of knowledge to experts occurs through the general division of labour, on the other hand the population, in part, lives more closely to a given reality in a given place. They are the most knowledgeable about their local affairs (as if to say 'laymen are the experts of everyday life.')"

Flüeler (2006b)

"...what we [expert anthropologists, tf] call our data are really our own constructions of other people's constructions of what they and their compatriots are up to .... The whole point of a semiotic approach to culture is ... to aid us in gaining access to the conceptual world in which our subjects live so that we can, in some extended sense of the term, converse with them."

Geertz (1973)

"Scientists and experts answer questions which by far surpass their own life expectancy."

A participant at an event of regional participation in the Swiss site-selection procedure, Zurich Northeast, 25 April 2013

And **time**. It needs time. The technical community needs to explain reasoned arguments plausibly to a lay public. It is not enough to build up static long-term records, knowledge and memory (Box 7.2, references) but to keep them and

Fig. 7.1 Proposal for (re-) structuring the institutional setting of radioactive waste governance —and respective others—in a state. Decisive is the control of resources ("Fund") and process ("National Council" as its "guardian"). C Committee, FD final disposal, HLW high-level waste, LLW low-level waste, MIR waste from medicine/industry/research, NGO non-governmental organisation, QA quality assurance, R&D research and development, SA safety analysis (adapted from Flüeler (2003b, 2006b), 273)

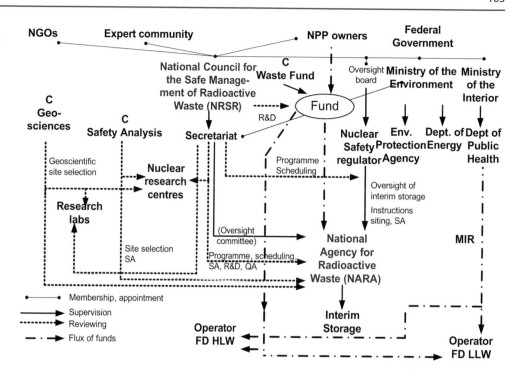

exchange live ("active heritage", cf. Pescatore and Palm 2020). Society, as well as policy, needs time to understand and tolerate the outcome (rather than accept it, because tolerance is all we can expect if waste is the topic).

7.2   Skills for Sustainable Governance

In order to truly engage in discourse and dialogue with conflicts as being normal, participants need to adopt and apply critical thinking (Box 7.3, References *Literacy in critical thinking and science*). It is a prerequisite of both science and problem-solving in general. Think of how Karl Popper defined it: "… the novelty of science and scientific method … is its consciously critical attitude to attempted solutions" (Popper 1999). It involves one key point also of safety culture, namely to have a questioning attitude and analyse my and other basic assumptions and background logic. It also contains a prudent sorting out of data and information to build a sound knowledge base. "The ideal scientist is a critical thinker who is humble enough to make the Confession of Ignorance, self-confident enough to admit it out loud, and courageous enough to set out on a course that will rectify the situation … even when others think there is no sense in it" (Carr 1995, 21).

A well-cultivated critical thinker

- raises vital questions and problems, formulating them clearly and precisely;
- gathers and assesses relevant information, using abstract ideas to interpret it effectively;
- comes to well-reasoned conclusions and solutions, testing them against relevant criteria and standards;
- thinks open mindedly within alternative systems of thought, recognizing and assessing, as needs be, their assumptions, implications and practical consequences;
- communicates effectively with others in figuring out solutions to complex problems.

https://www.criticalthinking.org/

Box 7.3: Critical Thinking

"The ability to think critically ... involves three things: (1) an attitude of being disposed to consider in a thoughtful way the problems and subjects that come within the range of one's experiences, (2) knowledge of the methods of logical inquiry and reasoning, and (3) some skill in applying those methods. Critical thinking calls for a persistent effort to examine any belief or supposed form of knowledge in the light of the evidence that supports it and the further conclusions to which it tends. It also generally requires ability to recognize problems, to find workable means for meeting those problems, to gather and marshal pertinent information, to recognize unstated assumptions and values, to comprehend and use language with accuracy, clarity, and discrimination, to interpret data, to appraise evidence and evaluate arguments, to recognize the existence (or non-existence) of logical relationships between propositions, to draw warranted conclusions and generalizations, to put to test the conclusions and generalizations at which one arrives, to reconstruct one's patterns of beliefs on the basis of wider experience, and to render accurate judgments about specific things and qualities in everyday life."

Edward M. Glaser (1941)

Critical thinking (Halpern 1989) alone does not ensure sustainable development (of regions, states or the planet). Competences thereof are as proposed by de Haan and colleagues in 2008 (adapted from Minsch et al. 2012a, 275; Sect. 7.5):

Interactive use of media and resources

- Adopting various perspectives: broad-based, new perspectives to support the generation of integrated knowledge;
- Anticipating situations: ability to analyse and assess forward-looking developments;
- Engaging in a cross-disciplinary approach: generating and using interdisciplinary insights;
- Dealing with incomplete or overly complex information: ability to recognise and weigh risks, dangers and uncertainties of various origin.

Interaction in heterogeneous groups

- Cooperating: ability to plan and implement actions with others;
- Coping with individual decision-making dilemmas: ability to deal with conflicting objectives by reflecting on strategies for action;

- Participating: ability to take part in collective decision-making processes;
- Being motivated: ability to motivate oneself and others to become active.

Independent activities

- Reflecting on general principles: ability to reflect on one's own and others' general principles;
- Engaging in moral actions: ability to incorporate ethical principles such as justice in decisions and actions;
- Acting independently: ability to plan and act independently;
- Supporting others: ability to show empathy for others.

In order to really become sustainable, our societies must become learning societies, with sufficient reflexivity beyond short-term linear thinking, mature and empowered citizens and anticipatory conflict management. Conflicts are normal. We need a climate for creativity, and a sense of responsibility and precaution for constructive handling of scarcities (Minsch and Flüeler 2012b).

7.3   "Organised Safety": Leading Role of Authorities

The alpha and omega of successful long-term programming is to keep all major actors on board, mainly the technoscientific community, waste producers, public administrators, NGOs, and the public, national and regional. As all three policy fields are driven by technology but decided upon by society, confidence in the system and trust in the responsible actors are central. Trust and confidence of the public depend on the "degree of organised safety" (Vlek and Stallén 1981), which was supposedly erected by the institutions involved. Since the expertise of external experts is necessary, the public has to gain trust in the scientific-technical community. Their judgement base, therefore, does not solely rest on expertise but is also, if not primarily, process-based. Consequently, not only is confidence in technical performance assessments needed but also trust in the persons and institutions in charge and participating in the chosen procedure. In complex technical domains, trust (in experts and their work) is a key notion in the transfer of knowledge. Particularly, when dealing with long-lasting waste, one cannot rely on known techniques (state of the art) but has to compensate for ignorance (i.e., here the absence of knowledge) by trusting in the specialised institutions (regulators, safety authorities, applicants and "independent" scientists). Main "safety organising" actors are the conductors of the procedures (e.g., the Department of Energy or of Environment), the regulatory and the supervisory authorities (e.g., the Federal Safety Inspectorate or the

Environmental Protection Agency) and the guardians of the process (e.g., a National Waste Council, cf. Minsch et al. below). Thereby, let us keep in mind the ambiguous role of the state, promoting, sometimes even implementing, certain industrial policies and having to regulate their systems at the same time. See Fig. 7.1.

The Nuclear Energy Agency describes the "Safety culture of an effective nuclear regulatory body" with five principles (NEA 2016, 8):

- "Leadership for safety is to be demonstrated at all levels in the regulatory body;
- All staff of the regulatory body has individual responsibility and accountability for exhibiting behaviours that set the standard for safety;
- The culture of the regulatory body promotes safety and facilitates co-operation and open communication;
- Implementing a holistic approach to safety is ensured by working in a systematic manner;
- Continuous improvement, learning, and self-assessment are encouraged at all levels in the organisation".

Consistent with this (and going well beyond), Minsch and colleagues proposed four groups of strategies in their "Institutional reforms for a policy of sustainability" (Minsch et al. 1998, see also Rychen and Salganik 2003, Parkin et al. 2004):

- Strategies of reflexivity with "systems of reporting", expert bodies, improved structuring of information in decision-making processes as well as research, science and education aimed at sustainability;
- Participatory strategies of self-organisation with rights to participate and discursive participatory models;
- Strategies to compensate and resolve conflicts with, e.g., "advocatory" institutions such as a "Council of Sustainability" established by the Government, "services of sustainability" in organisations, free access to information, compensation for NGOs and controls of monopoly, "better integration of NGOs… into negotiation processes";
- Innovation strategies with, i.a., a legal liability system, and a "co-operative development of the regional level".

Due to the driving force of technology (waste "exists" and "seeks" a "solution" whatever that might be), the entire sociotechnical system is heavily predetermined. Being stakeholders themselves, the regulators have to establish a platform for inclusive knowledge generation. This necessity to integrate different requirements, the step-by-step approach, the chance of "institutional constancy" and the special "national" task of the issue call special attention to the role of the regulatory authorities (NEA 2003, 2004a, b, 2012). Box 7.4 illustrates how sensitive the issue is.

Box 7.4: Democracy Suffers When Government Statistics Fail

"The Web was invented 30 years ago, yet 2020 is the first year that the United States' decennial census has allowed households to respond online. This shift came in the nick of time—given the need for social distancing—and has been a (qualified) success ….

The lack of innovation in federal statistical agencies such as the Census Bureau contributes to the slow, bureaucratic and hidebound nature of government, argues New York University economist Julia Lane in *Democratizing Our Data: A Manifesto*. Many factors are to blame: a dearth of modern technology, absence of adequate training in new data-science techniques such as machine learning, outdated legal rules that preserve inflexible practices, and budgetary processes that create silos and impede collaboration ….

Until recently, the US federal government spent billions on science and technology research with no idea about the return on investment. Moreover, we lack skilled people in government to make use of this rich data store to inform how policy is made. Given that the average salary for a senior data scientist in Silicon Valley is twice that of a senior civil servant, this is no wonder ….

Paraphrasing Robert F. Kennedy, Lane writes about [the Gross Domestic Product, GDP]: 'We judge the United States by production—we count air pollution, cigarette advertising, locks for our doors and jails for people who break them. We count the destruction of the redwood, the production of nuclear warheads, but not the health of our children, the beauty of poetry, the intelligence of our public debate or the integrity of our officials …. [GDP] measures everything, in short, except that which makes life worthwhile.' While we may not want to measure the beauty of poetry, her point is that we are stuck without the ability to experiment with new forms of measurement ….

Lane argues that Congress should create a 'National Lab for Community Data' that operates outside the government. Like other federally funded national research labs, such as Lawrence Livermore in California, which accelerate important research in the public interest, it would enjoy access to well-trained talent from outside, and data from inside, government. And because it would be quasi-external, it would be able to innovate faster and be more responsive to citizens' needs, leading to the creation of more relevant data ….

… Given … the broader national debate around the politicization of science agencies during the COVID-19 epidemic, more discussion about politics in *Democratizing Our Data* might bolster Lane's argument for external collaboration. Also welcome would have been a longer discussion about statistical agencies in other countries—what is working and what is not around the world—as well as further exploration of how to engage the public in these important debates.

For now, this pithy volume is a must-read account of what the US federal statistical agencies are, what they do and why public statistics are vital to democracy. If we cannot be counted, we cannot be heard."

Beth S. Noveck (2020)

7.4 The Road: Neither a Supreme Algorithm nor a Cookbook

Hardware does not replace software. Structures are necessary but do not substitute engagement and morale. This book is a thought-and-action provoker. We cannot delegate the issues (even though we will have to, considering the timescales), neither to machines nor to the Panel of the Wise (not even the National Council proposed). There is no golden standard nor a sophisticated formalism (e.g., Diaz and Ewing 2018), let alone an attempt at a universal (governance) theory. Because of the complexities of the system/programme this is not conceivable. Its reason is not any type of ordinary uncertainty but its inherent indeterminacy. Even though we may—and should—do our best, we have no guarantee that our children's children (experts, politicians, companies, publics ...) will follow us. Memory and records are fine (see NEA's initiative, NEA 2019, *Archives, Memory,* etc. in references). This is commendable but insufficient. The best archives are static (Box 7.2); to achieve enough drive, to actually obtain programme understanding and a transfer of knowledge through time, needs adaptive dynamics of all involved. What we need is a societal thrust! The skills demanded by the challenges, the major role of process aspects, the clash of perspectives and the longevity of the undertakings highlight the importance of educated people. The respective literacy needed is a major indication that education is system-relevant. In analogy to the WHO Director General's dictum with regard to health and the COVID-19 pandemic (WHO 2020b), education is not a cost—it is an investment. And I would add: a necessary one, and one that gives pleasure at that.

7.5 Open Frontier: Tools and Roads to Explore

Having stated several reservations, of course, there are plenty of avenues to go as this book is an annotated overview of the vast subject and also somewhat a call for

proposals. Reaching beyond the conceptual outline, some areas might be worthwhile considering against the backdrop of strategic decision making (Sects. 6.1 and 6.2, coined in Fig. 6.5). In view of the wide range of disciplinary fields involved and the major characteristic of the applications, namely indeterminacy, it is not surprising that now we cannot list a full-fledged and specific research agenda but suggest a few pertinent roads to explore concerning problem identification (Sects. 7.5.1, 7.5.2, 7.5.4), problem development and selection (7.5.3) and feedback (7.5.5, 7.5.6).

7.5.1 Knowledge Gathering: Platforms, etc.

As decision problems are informational problems and information is incomplete in principle, we have to be keen on finding/selecting the "adequate" information needed to enter a comprehensive discourse and eventually reach an informed decision—as "objective" and unbiased as possible. It is a useful idea to set up an **open online platform**. Refer, as a current example, to the UN COVID-19 initiatives, partnerships and activities (ITU 2020) with their "Global Network Resiliency Platform. Best practices to improve COVID-19 responses" (GNRP 2022) provide country-specific information. Evidently, many targeted information packages are available, some technical (IGDTP 2022; Sitex 2022), others more comprehensive, such as on the Nuclear Energy Agency's website regarding radioactive waste management, e.g., specifically on its societal aspects (NEA 2022, web) even though some might object that it would be a partisan page. Many integrative research projects have evolved into platforms (Fig. 6.4) or complementary research to enrich the database including lessons learned. Whether the platform can develop to a **strategic (digital) dashboard**, e.g., to visually track key performance indicators, remains to be seen.

With respect to a pluralistic, yet not antagonistic, discourse, a polarisation as in the Coronavirus debate (with two expert community positions[3]) should be avoided. Instead of blocking dissident blogs on social media, a good practice is to **table and respond to controversial views and assertions** as is being done by WHO in the case of the pandemic, or by climate professionals in the climate change controversy (WHO 2020; Skeptical Science Team 2022).

---

[3] "Great Barrington Declaration" (2020, 4 Oct 2020) versus the "John Snow Memorandum" issued by mainstream epidemiology (2020, 15 Oct 2020). Note: this is not the case in climate science even though there are minority views. See Cook et al. (2016) and http://theconsensusproject.com/.

7.5.2 From Participation to Open Discourse: Focus Groups 3.0., etc.

The three-step approach in Section 3.4 foresees a comprehensive societal discourse. This poses high challenges as experience shows[4] and research acknowledges (e.g., InSotec 2014). To understand and assess, not to decide on, the present basic assumption on who should participate in the long-term processes under scrutiny is that all relevant perspectives should be on board, not as many as possible. The majority rule may apply when it comes to votes or vetos on national or regional levels. Remember what a "socially robust" decision would encompass: most arguments, evidence, social alignments, interests and cultural values would lead to a consistent option. If social alignments, interests and values are to be sincerely incorporated "participation" has to be, at least, "collaborating", if not "empowering" (Fig. 7.2). There is a vast literature on techniques (see *Tools* references) but the **paramount questions** which must be addressed openly are: **What is participation for, with whom and at what aim?** The techniques to use, such as consensus conferences, focus groups or multi-criteria mapping must be thoroughly assessed and targeted to the expected use (Flüeler et al. 2007f; Krütli et al. 2010b; Stauffacher et al. 2012). One thing is for sure: Objectives for, and limits to, involvement have to be defined at the outset of the process so that all participants are aware of the scope and can decide accordingly. What the physician and science theorist Ludwik Fleck developed for scientific thinking almost a century ago (Box 7.5) also applies to sociotechnical thinking and problem-solving: Though we will certainly not equalise our "thought structures", we may reach some minimal "thought collective" to carry on and eventually out of what we have left behind, a sustainable governance of long-term waste.

All participatory activities are judged against the backdrop of a "fair" procedure and treatment of the intra- and intergenerational equity issues, taking into account the twofold—spatial and temporal—asymmetry: the benefit of nuclear electricity is broadly distributed (today), whereas eventual costs or risks of waste disposal are locally concentrated and transferred to future generations.

Box 7.5: Introduction to Thought Collectives

A thought collective exists wherever two or more people are actually exchanging thoughts. He is a poor observer who does not notice that a stimulating conversation between two persons soon creates a condition in which each utters thoughts he would not have been able to produce either by himself or in different company. A special mood arises, which would not otherwise affect either partner of the conversation but almost always returns whenever these persons meet again. Prolonged duration of this state produces, from common understanding and mutual misunderstanding, a thought structure [Denkgebilde] that belongs to neither of them alone but nevertheless is not at all without meaning. Who is its carrier and who its originator? It is neither more nor less than the small collective of two persons. If a third person joins in, a new collective arises. The previous mood will dissolve and with it the special creative force of the former small collective ....

An 'ultimate' or set of fundamental first principles from which such findings could be logically constructed is just as non-existent as this 'everything.' Knowledge, after all, does not repose upon some substratum. Only through continual movement and interaction can that drive be maintained which yields ideas and truths.

Ludwik Fleck (1935)

7.5.3 Outcome: MCDA 3.0. td Labs, etc.

There are many ways to reach an outcome—by vote, negotiation, calculation, etc. One method is the **multi-criteria decision analysis, MCDA**. This is not the place to portray the wide and fast-growing field of decision-making toolkits (e.g., IAEA 2019; Saarikoski et al. 2016; Yap et al. 2019). Essential is to say that it is not designed to provide the best options but to find a suitable trade-off among options for given circumstances; which technique to use depends on the problem context and the information available. Most challenging and worth more research is the **integration of data, information, knowledge and values/views**. For example, there are several levels of conflict, among "factual" (judgement) and "value" conflicts, such as on data and statistics, estimates and probabilities, assumptions and definitions, risk-cost-benefit trade-offs, the distribution of risks, costs and benefits and basic social values (von Winterfeld and Edwards 1984). MCDA attempts to systematically consider and reveal stakeholder preferences (KIT 2020). As we are dealing with contentious and overly complex issues, we should explore all types of paths to tackle them, like policy labs, citizen labs, transdisciplinary labs, etc. (Box 7.6).

---

[4] Radioactive waste: initially positive, then mixed in Canada (NWMO 2005, 2022; Yellowhead Institute 2020; CBC 2020), failed as "débat public" in France (CNDP 2022), cautiously positive in Switzerland (SFOE 2020), straightforward targeted on siting - and long-term nuclear - community in Finland and Sweden (WNWR 2019; Sweden 2022). CCS: See Upham and Roberts (2011), Feenstra et al. (2010).

Fig. 7.2 "Participation" can mean a lot of things—even manipulation[5] (to the left, not on the figure) (reproduced from © IAP 2018, 2022)

	INFORM	CONSULT	INVOLVE	COLLABORATE	EMPOWER
PUBLIC PARTICIPATION GOAL	To provide the public with balanced and objective information to assist them in understanding the problem, alternatives, opportunities and/or solutions.	To obtain public feedback on analysis, alternatives and/or decisions.	To work directly with the public throughout the process to ensure that public concerns and aspirations are consistently understood and considered.	To partner with the public in each aspect of the decision including the development of alternatives and the identification of the preferred solution.	To place final decision making in the hands of the public.
PROMISE TO THE PUBLIC	We will keep you informed.	We will keep you informed, listen to and acknowledge concerns and aspirations, and provide feedback on how public input influenced the decision.	We will work with you to ensure that your concerns and aspirations are directly reflected in the alternatives developed and provide feedback on how public input influenced the decision.	We will look to you for advice and innovation in formulating solutions and incorporate your advice and recommendations into the decisions to the maximum extent possible.	We will implement what you decide.

INCREASING IMPACT ON THE DECISION →

Box 7.6: Transdisciplinary Labs

"Transdisciplinary case studies (tdCS) are problem-oriented, research-based seminars held in real-life situations, to facilitate the production of scientific and praxis-related joint knowledge. In a tdCS, students from a diverse set of academic backgrounds approach a real-world sustainability problem (e.g., [solid waste management]). They collaborate with stakeholders and experts in and outside academia to bridge an intensive exchange between academia and practise. Throughout the tdCS, interdisciplinary methods are conducted, such as scenario analysis, sustainability assessment, and multi-criteria assessment, which are coupled with interdisciplinary methods like material flow analysis, cost benefit analysis, in-depth interviews, and surveys."

Adelene Lai et al. (2016)

7.5.4 Handling Uncertainties: Addressing Indeterminacy

Complex systems are characterised by non-linear interactions, chaotic dynamics, the "butterfly effect", phase transitions, self-organised criticality, cascading effects and power laws.

Instead of controlling such systems, it may be useful to guide their self-organisation, i.e., "to use the driving forces of the system rather than to fight against them" (Helbling 2009). You would not turn the helm abruptly to change the course of a sailing boat or turn the steering wheel hard when your car is sliding on ice. If a storm arises, you would reef or even drop sails; on ice you would drive on the road shoulder to slow down.

Uncertainty might be handled likewise—instead of calculating more runs with a sophisticated sensitivity analysis one would rather follow the drive of the system and steer it cautiously. For we are confronted with **indeterminacy, not "just" uncertainty**, even of the major kind such as epistemic uncertainty or ambiguity (Hester et al. 2003; Flüeler 2006b, 144ff). It does not need wars or economic breakdowns (Fig. 5.4c/d) for latency phases to emerge, when, e.g., the state is confronted with a new task or an old problem arises in a new fashion, in which the political arena is newly set, where the stakeholders have not yet taken up position (or abandoned it) and where the rules of the game are not fixed yet (Freiburghaus and Zimmermann 1985, 88f). Our sights may have to be lowered as each generation will have the same task (to carry on the respective programme), in order to instil and have people get a sense of ownership of the problem, on all levels (from the technical community to the local community), today. Any long-term solution to a sociotechnical problem must start with the short term—with the idea of the rolling present as the first step to the future. This may be a road to minimise indeterminacy. Others may

---

[5] Mentioned by Arnstein in her seminal "participation ladder" (Arnstein 1969).

be to operationalise productive and forward-looking concepts such as resilience (Kovalenko and Sornette 2013).

7.5.5 Evaluation, Effectiveness, Efficiency: Before, During, After

Traditionally, evaluation is not included in decision making even though it is essential for mutual institutional and social learning. Given the time dimension, policy evaluation is a clear requisite to investigate the efficiency and effectiveness of the programmes in question, in order to examine their content, implementation, or impact.[6] Ideally, it consists of three phases: *ex ante* (before the programme is launched, in the planning), ongoing (during the course, also to adapt in case of need) and *ex post* (when the programme is completed). **Monitoring and evaluation** go hand in hand, whereby the former "is checking progress against plans" and the latter (traditionally) is assessing terminated projects, programmes or phases thereof (sportanddev.org 2022, HM Treasury 2020, 6). It is primal to involve multiple stakeholder groups bringing multiple perspectives to the table—otherwise, they will attract attention on some other channel anyway.

However sophisticated the evaluation technique may be, it has to adequately address the following questions (OECD 2022) whereby "intervention" means "programmes" in our cases:

- Relevance: Is the intervention doing the right things?
- Coherence: How well does the intervention fit?
- Effectiveness: Is the intervention achieving its objectives?
- Efficiency: How well are resources being used?
- Impact: What difference does the intervention make?
- Sustainability: Will the benefits last?

With regard to effectiveness "versus" efficiency, we may keep our definition of "good decisions" and Gross Stein's dictum in mind: "Without discussion of purpose, effectiveness makes no sense, and without discussion of effectiveness, efficiency is stripped of its analytic power and becomes a cult" (Gross Stein 2002, 192).

7.5.6 Root-Cause Analysis

The best way to treat waste is not to produce it at all. Consequently, it is useful and demonstrates a truly

---

[6] In view of the duration of the programmes, I do not differentiate between policy and programme evaluation (as opposed to CDC 2022). See also Guba and Lincoln 1989, Lay and Papadopoulos 2007, Interact 2022.

integrative understanding to go back to the roots. Remember the starting point: The waste community agrees that waste is either a resource in the wrong place (not yet recycled) or a residual material to be disposed of (Sect. 4.4).

In the **nuclear field**, the majority of current waste stems from its generation in nuclear power plants. Reprocessing was long seen as closing the nuclear cycle by extracting plutonium and uranium and reusing these valuables in fresh fuel. If reprocessing were resumed and brought to a higher environmental standard than today's facilities provide, followed by a separation and transmutation of major radionuclides on an industrial scale (not just in the lab as currently), nuclear fission would be on a sustainable path. Another one might be to promote and roll out advanced reactor technologies (Sornette et al. 2019, 199ff). This presupposes a strong and dedicated will to massively invest in research and development, R&D, convince society and implement the nuclear fleet.

Waste is "the mother of all environmental problems" (Kunzig 2020). A fully sustainable resource policy assumes as **closed material cycles** as possible and thus, just interim storage—not disposal—for future resource use (EC 2018). Within its Circular Economy Package, the European Union plans to phase out landfilling for recyclable waste (including plastics, paper, metals, glass and biowaste) in non-hazardous waste landfills by 2025, corresponding to a maximum landfilling rate of 25% (EU 2019). If we are determined to follow this route, the statement would come true: "The resources we need are no longer in the ground, but in landfill" (Treggiden 2020). Remember again (Sect. 5.6), by throwing away electronic devices alone, we currently lose 57 billion USD of secondary raw material as many goods contain valuable substances. A tonne of mobile phone waste alone is worth 10,000 EUR (Forti et al. 2020).

Carbon capture and storage, CCS, is usually conceived as a mitigation strategy to gain time for a transition to fossil-free energy systems, it may also serve to produce hydrogen and foster a negative emissions path with bioenergy (Sect. 4.5). **CCUS** (including Utilisation) would take it from the stigma of a "quick fix" (and trick) by the fossil industry and, instead, lend a flavour of proactive, forward-looking technology of the future. This, again, requires dedicated policies on the international, national and regional levels with a provision of significant resources (Sect. 2.8).

7.6 Above All it Needs Engagement, Respect and Trust

Actors (guardian, experts of various origin, publics), structures, resources and novel tools are fine but only if people are willing to enter a dialogue, sometimes reaching across the aisle, finding a consensus on the goal(s) and following a process with clear rules. A good product is indispensable (a safe and safely run repository in a suitable site), but the process is also important—to stakeholders and the public even pivotal. And why so? Because this is what lay people can assess and where experts can show their true competence and politicians their truthfulness and accountability. In addition, it is the way we treat each other right now that will be the yardstick for our grandchildren to judge what we are doing—and if they will be willing to take over (Fig. 7.3).

a Do what your boss tells you (Photo: media conference of TEPCO)

b Don't bother – limit your involvement to protesting against restrictions of your civil liberties (Photo: rally against Corona pandemic measures, AP, September 2020)

Fig. 7.3 We are given the choice which scenario to follow (Photos by **a** Tepco 2011, **b** AP 2020, **c** SFOE 2008; McDonnell 2020)

c Engage – help decide and establish informed consent (Photos: participation in view of Swiss site-selection process; line of voters for US presidential election, September 2020)

Fig. 7.3 (continued)

Apart from technical competence, experts and agency staff need an adequately developed culture, respect for others, admitting failures, stamina and flexibility, change of perspective and some empathy. Stalemates and in-fighting wear people out, reluctancy rubs off on other discussants and indignation may have the opposite effect of what was intended. We cannot leave this issue to "heritage managers" or "futurologists", especially knowing that there is no such particular skill (Högberg et al. 2018).

Lay people concentrate on the process and on actors whom they are very well able to judge.

– Were the rules complied with?
– Are the experts credible, even authentic?
– Are they arrogant?
– Do they admit mistakes?
– Do they really address (my) questions/remarks?

The public's indicator, therefore, is confidence in the process and trust in the personnel.

7.7   Overall Conclusions

Dealing with persistent complex sociotechnical systems—and long-term energy regimes are such—requires acknowledging some basics. It

- needs an integrated perspective. It is not sufficient to "solve" subtopics and to subsequently add them up;
- is "transscientific" in nature. We are reminded of Weinberg's dictum that questions "which cannot be answered by science" are "trans-scientific" (Weinberg 1972);
- can only be decided on by society. This does not diminish the role of experts, science and research—to the contrary, they are more challenged than ever in the sense that the issue
- is transdisciplinary. Such research goes as far as "to make the change *from research for society to research with society*" (Scholz 2000, 13);
- is often a transgenerational issue (the risk situations and project management mentioned throughout the book);
- has to be transpolitical (to overcome ephemeral politics such as NIMTOO effects = Not In My Term Of Office).

To consider both technical and social issues needs an inclusive, systematic and participatory approach to single out goal priorities (presumably with safety first). Setting up a respective process is a prerequisite to proceed with site selection (and further programme steps). It is essential to have a (national) lead agency in conjunction with a clear division of roles among the players, rules of the "game" and criteria to judge. In addition, a proposed oversight body surveils the programme and its focused implementation. The complex and long-lasting procedure necessitates extensive resources on all sides and of all types over time.

The value of a dialogue, above all, depends
on the diversity of controversial opinions.
No rational argument has a rational effect on somebody
who does not want to assume a rational attitude.

Karl Popper 1982, The open universe. An argument for indeterminism

Our society's success in credibly addressing intragenerational issues might convince future generations to be willing to carry on the programmes when needed. According to the concept of social robustness, the concerned and deciding stakeholders must achieve consent on some common interests, at least on three levels: problem recognition, consensus on the main goals and the procedural strategy (Sect. 3.3). As to knowledge transfer, the challenge is to ensure a continual process so that the broadly consented goals can be understood, agreed to and followed by generations to come. Goal priorities must be set, presumably with safety first but embedded in the affected siting region. Respective processes must be set up to proceed in site selection (and subsequent steps). This requires appropriate structures, broadly supported leadership, accountable and trustworthy implementers and authorities as well as engaged people with a sense of ownership of the problem on the local/regional level.

It has thus become clear that the institutional aspects are more and more getting to be the linchpin of the issue and maybe of the solution:

- The long-lasting and entwined project character rests on the constancy of competent and trusted institutions;
- Society may only exert indirect control on such complex technological projects as the one at hand, via institutional paths (e.g., Kasperson et al. 1982; Fischhoff 1977; Thomas et al. 1980); the main quality check in science, at that, is institutional peer reviewing (Jasanoff 1985);
- The public appraises technologies, thus nuclear, as a whole, including the respective institutions (Wynne 1980) and their achieved "degree of safety" (Vlek and Stallén 1981);
- The debate on risks is also a debate on democracy and progress, it is sparked off by the "controversy over the institutionalisation and regulation of the progress of technological knowledge" (Evers and Nowotny 1989, 247); Kasperson and colleagues went so far as to coin the "risk crisis" to be truly an "institutional crisis" (Kasperson et al. 1992) and, in consequence a "regulatory crisis" (Hutter and Lloyd-Bostock 2017). As "crisis" is misleading in our cases and as we need a long view, the focus must shift from a demand-and-control and management mentality to an approach of more resilience. We need a broad societal dialogue, with an "educated public" (Brown 2016), together with willing and authentic experts so that, by mutual learning, we truly reach more literacy in the long-term governance of the three policy fields under scrutiny. This poses high demands on a society with its institutions, especially in countries with weak structures.

The broader the societal agreement on key issues is (e.g., what is the main goal of a programme, what are complementary goals? Where is consensus, where dissent, where compromise? How safe is safe enough? When shall monitoring be terminated, and on what grounds?) the more valuable—"robust"—and useful is the social pool (Gelobter 2001), and, at that, also the technological, resource the future generations can draw from. Strategic Monitoring may serve to cover the mentioned aspects in radioactive waste, conventional hazardous waste and CCS/CCUS governance.

Strategic Monitoring is devised to provide process, implementation or policy and institutional surveillance to sustain a once launched programme. It includes the tailored transfer of knowledge, concept and system understanding, experience and documentation to specific (involved and responsible) audiences: the technoscientific community, the waste producers and implementers, the public administrators in charge, NGOs and the public. It not only addresses the controversial long-lasting "problem" (of nuclear, toxic or CO_2 waste) but also investigates ways to approach solution spaces, not just technical but also institutional and personal, and this for the long term. It is an integrative tool of targeted yet adaptive management and may be applicable to other long-term sociotechnical fields.

It is still noteworthy, and—with modesty—valid, what Parker and colleagues reflected almost four decades ago on the disposal of high-level radioactive waste (inverted commas in the original):

> Because of the equivocal and subjective nature of the standards, people with different value systems will derive very different conclusions about the adequacy of the solution. Therefore, because of the varied value systems in democratic societies, it is obvious that total consensus cannot be reached. Consequently, the only valid course is to adopt a process that is as open, complete and fair as possible, and to try to develop as broad a consensus as possible in support of the solution that is finally reached. It has to be recognized that there will always be an irreducible amount of uncertainty in the outcome of any solution …. It is apparent that there is no rigorous, scientifically valid solution to the high-level radioactive waste disposal problem until we define what a "solution" is. It is obvious that the definition is not scientific alone. Consequently, the "solution" in every country will be distinctive and unique to that country. The Swedish feasibility study (KBS-3) has been compared to "solutions" found or sought in a number of other countries …. As mentioned so frequently in this paper, a "solution" is to be found only in a national, not a scientific, context.
>
> Frank L. Parker et al. (1984), 101, 108

τόλμα πρήξιος ἀρχή, τύχη δὲ τέλεος κυρίη.
Courage is the beginning of an action,
but chance is the master of the end.

Democritos, Greek philosopher (460–370 BC), Fragment 269

Wer immer strebend sich bemüht/Den können wir erlösen;
Who strives always to the utmost, For him there is salvation;

Johann Wolfgang von Goethe (1749–1832), Faust II,
Act 5, verse 11936f

Exercise 7.1

Discuss the idea of a guardian of the programme and process (Fig. 7.1). How would it have to be installed to be successful?

Exercise 7.2

Use Fig. 7.1: Analyse countries how the polluter-pays principle is applied and compare the respective regulators and implementers.

Additional Information

All weblinks accessed 27 January 2023.

Key Readings

Key References

Flüeler T (2002f) Long-term radioactive waste management: challenges and approaches to regulatory decision making. HSK/IAEA/NEA workshop on regulatory decision making processes. Grandhotel Giessbach, Brienz, Switzerland. 15–18 Oct 2002

Flüeler T (2004a) Long-term radioactive waste management: challenges and approaches to regulatory decision making. In: Spitzer C, Schmocker U, Dang VN (eds) Probabilistic safety assessment and management 2004. PSAM 7–ESREL '04. Berlin, June 14–18, vol 5, Springer, London, pp 2591–2596. https://link.springer.com/chapter/10.1007%2F978-0-85729-410-4_415

Flüeler T (2006b) Decision making for complex socio-technical systems. Robustness from lessons learned in long-term radioactive waste governance. Environment & Policy, vol 42. Springer, Dordrecht NL, pp 269ff. https://doi.org/10.1007/1-4020-3529-2_9

Normality

Asonye C (2020) There's nothing new about the 'new normal'. Here's why. The World Economic Forum COVID action. 5 June 2020. https://www.weforum.org/agenda/2020/06/theres-nothing-new-about-this-new-normal-heres-why/

Prêtre S, Lochard J (1995) Return to normality after a radiological emergency. Health Phys 68(1):21–26

Reinhart CM, Rogoff KS (2009) This time is different. Eight centuries of financial folly. Princeton University Press, Princeton, NJ, 512 pp

Literacy in Critical Thinking and Science

Carr JJ (1995) The art of science. A practical guide to experiments, observations, and handling data. HighText publications, San Diego, CA

Clarke J (2019) Critical dialogues: thinking together in turbulent times. Policy Press, Bristol UK

Critical Thinking (2022, web) Foundation for critical thinking. http://www.criticalthinking.org/pages/our-conception-of-critical-thinking/411, https://www.criticalthinking.org/pages/defining-critical-thinking/766

Davies M, Barnett R (2015) The Palgrave handbook of critical thinking in higher education. Palgrave Macmillan, New York

Fleck L (1935, 1979) Genesis and development of a scientific fact. Trenn TJ, Merton RK (eds), transl. Bradley F, Trenn TJ. University of Chicago Press, Chicago, IL. Orig.: Fleck L (1935, 1980) Entstehung und Entwicklung einer wissenschaftlichen Tatsache. Einführung in die Lehre vom Denkstil und Denkkollektiv. In: Schäfer L, Schnelle T (eds). Suhrkamp Taschenbuch Wissenschaft, Frankfurt aM

Glaser EM (1941) An experiment in the development of critical thinking. Teacher's College, Columbia University, New York. https://www.criticalthinking.org/pages/defining-critical-thinking/766

Halpern DF (1989) Thought and knowledge: an introduction to critical thinking. Lawrence Erlbaum, Hillsdale, NJ

Jasanoff S (1985) Peer review in the regulatory process. Sci Technol Human Values 10(3):20–32

Popper K (1999) All life is problem solving. Therein: the logic and evolution of scientific theory (1972). Routledge, Abingdon, Oxon, UK

Hester RE, Harrison, RM, Stirling A (2003) Renewables, sustainability and precaution: beyond environmental cost-benefit and risk analysis. Issues Environ Sci Technol 19. https://doi.org/10.1039/9781847551986-00113

Weinberg AM (1972) Science and trans-science. Minerva 10(209–222):209

Literacy in Sustainable Development

Brown GE Jr (2016) Nuclear waste disposal, climate change, and Brexit: the importance of an educated public. Elements 12(4):227–228. https://doi.org/10.2113/gselements.12.4.227

de Haan G, Kamp G, Lerch A, Martignon L, Müller-Christ G, Nutzinger HG (eds) (2008) Nachhaltigkeit und Gerechtigkeit. Grundlagen und schulpraktische Konsequenzen [Sustainability and justice. Fundamentals and consequences for practical teaching]. Ethics of Science and Technology Assessment 33. Springer, Berlin

Fischhoff B (1977) Cost benefit analysis and the art of motorcycle maintenance. Policy Sci 8:177–202

Gelobter M (2001) Integrating scale and social justice in the commons. In: Burger J (ed) Protecting the commons. A framework for resource management in the Americas. Island Press, Washington, DC, pp 293–326

Kasperson R, Hohenemser C, Kasperson JX, Kates RW (1982) Institutional responses to different perceptions of risk. In: Sills DL, Wolf CP, Shelanski VB (eds) Accident at Three Mile Island: the human dimensions. Westview Press, Boulder, CO, pp 39–46

Minsch J, Flüeler T, Goldblatt DL, Spreng D (2012a) Lessons for problem-solving energy research in the social sciences (Chap. 14). In: Spreng D, Flüeler T, Goldblatt D, Minsch J (eds) Tackling long-term global energy problems: the contribution of social science. Environment & Policy, vol 52. Springer, Dordrecht NL, pp 273–319

Minsch J, Flüeler T (2012b) Die Schweiz, ein immerwährendes Gespräch. Multiple Krise, Komplexität, Unsicherheit, Konflikte: Anforderungen an eine Transformation zu einer Nachhaltigen Entwicklung & die Rolle von Wissenschaft und Bildung [Switzerland, an ongoing discourse. Multiple crises, complexity, uncertainty, conflicts: requirements for a transformation to sustainable development & the role of science and education]. In: Stiftung Zukunftsrat (ed) Haushalten & Wirtschaften. Bausteine für eine zukunftsfähige Wirtschafts- und Geldordnung. Zürich/Chur, Rüegger, pp 23–27

Minsch J, Feindt PH, Meister HP, Schneidewind U, Schulz T (1998) Institutionelle Reformen für eine Politik der Nachhaltigkeit [Institutional reforms for a policy of sustainability]. Springer, Berlin

Parkin S, Johnston A, Buckland H, Brookes F, White E (2004) Learning and skills for sustainable development. Developing a sustainability literate society. Guidance for Higher Education Institutions. Forum for the Future, London

Rychen DS, Salganik LH (eds) (2003) Key competencies for a successful life and wellfunctioning society. Hogrefe und Huber, Cambridge, MA

Scholz RW (2000) Mutual learning as a basic principle of transdisciplinarity. In: Scholz RW, Häberli R, Bill A, Welti M (eds) Transdisciplinarity: joint problem-solving among science, technology and society. Proceedings of the International Transdisciplinarity 2000 Conference, Workbook II: Mutual learning sessions 2. Haffmanns Sachbuch, Zürich, pp 13–17

Thomas K, Swaton E, Fishbein M, Otway HJ (1980) Nuclear energy: the accuracy of policy makers' perceptions of public beliefs. Behav Sci 25:332–344

Regulator's Role, Organised Safety

Hutter BM, Lloyd-Bostock S (2017) Regulatory crisis. Negotiating the consequences of risk, disasters and crises. Cambridge University Press: Cambridge. https://doi.org/10.1017/9781316848012

NEA, Nuclear Energy Agency (2003) The regulator's evolving role and image in radioactive waste management. Lessons learnt within the NEA Forum on Stakeholder Confidence. No. 4428. OECD, Paris

NEA (2004a) Learning and adapting to societal requirements for radioactive waste management—key findings and experience of the Forum on Stakeholder Confidence. OECD, Paris

NEA (2004b) Stepwise approach to decision making for long-term radioactive waste management. Experience, issues and guiding principles. No. 4429. OECD, Paris

NEA (2012) The evolving role and image of the regulator in radioactive waste management. Experience, issues and guiding principles. Trends over two decades. No. 7083. OECD, Paris

NEA (2016) The safety culture of an effective nuclear regulatory body. No. 7247. OECD, Paris. https://www.oecd-nea.org/upload/docs/application/pdf/2019-12/7247-scrb2016.pdf

Vlek C, Stallén PJ (1981) Judging risks and benefits in the small and in the large. Organ Behav Hum Perform 28:235–271

Archives, Memory, Etc.

Buser M (2013) A literature survey on markers and memory preservation for deep geological repositories. Mai 2010. OECD, Paris. https://www.oecd-nea.org/jcms/pl_19357/a-literature-survey-on-markers-and-memory-preservation-for-deep-geological-repositories?details=true

Högberg A, Holtorf C, May S, Wollentz G (2018) No future in archaeological heritage management? World Archaeol 49(5). https://doi.org/10.1080/00438243.2017.1406398

Holtorf C, Högberg A (2021) (eds) Cultural heritage and the future. Routledge, London, New York

Joyce R (2020) The future of nuclear waste. What art and archaeology can tell us about securing the world's most hazardous material. Oxford University Press, Oxford

Kahn A (2020) Archives of the planet. https://www.opendatasoft.com/blog/2016/07/22/archives-of-the-planet-albert-kahn-open-data

Kaplan MF (1982) Archaeological data as a basis for repository marker design. BMI/ONWI-354. Office of Nuclear Waste Isolation, ONWI, Battelle Memorial Institute, Columbus, OH

NEA, Nuclear Energy Agency (2014) Preservation of records, knowledge and memory across generations. Monitoring of geological disposal facilities—technical and societal aspects. OECD, Paris

NEA (2019) Preservation of records, knowledge and memory (RK&M) across generations. Final report of the RK&M Initiative. No. 7421. OECD, Paris

Otlet P, La Fontaine H (2020) Mundaneum. http://www.mundaneum.org/en

Pescatore C, Palm J (2020) Preserving memory and information on heritage and on unwanted legacies—new tools for identifying sustainable strategies to prepare and support decision making by future generations. SCEaR Newsletter 2020/1. UNESCO Memory of the World Programme. Sub-Committee on Education and Research (SCEaR), pp 4–14

Sebeok T (1984) Communication measures to bridge ten millennia. BMI/ONWI-532. Office of Nuclear Waste Isolation, ONWI, Battelle Memorial Institute, Columbus, OH

SKB (1996) Information, conservation and retrieval. SKB Technical Report 96–18. Swedish Nuclear Fuel and Waste Management Company SKB, Stockholm

Tannenbaum PH (1984) Communication across 300 generations: deterring human interference with waste deposit sites. BMI/ONWI-535. Office of Nuclear Waste Isolation, ONWI, Battelle Memorial Institute, Columbus, OH

Tools (and Roads to Go)

Arnstein SR (1969) A ladder of citizen participation. J Am Inst Plann 35(4):216–224

CDC, National Center for Injury Prevention and Control (2022, web) Brief 1: overview of policy evaluation. https://www.cdc.gov/eval/

Cook J et al (2016) Consensus on consensus: a synthesis of consensus estimates on human-caused global warming. Environ Res Lett 11:048002. https://doi.org/10.1088/1748-9326/11/4/048002

Diaz-Maurin F, Ewing RC (2018) Mission impossible? Socio-technical integration of nuclear waste geological disposal systems. Sustainability 10(12):4390, pp 39. https://doi.org/10.3390/su10124390

EC, European Commission (2018) Communication from the Commission to the European Parliament, the Council, the European Economic and Social Committee and the Committee of the Regions on the implementation of the Circular Economy Package: options to address the interface between chemical, product and waste legislation. Com(2018)32 final. EC, Strasbourg

EU, European Union (2019) Review of waste policy and legislation. https://ec.europa.eu/environment/waste/target_review.htm

Flüeler T, Krütli P, Stauffacher, M (2007f) Tools for local stakeholders in radioactive waste governance/long version. Challenges and benefits of selected PTA techniques. PTA-1. Work Package 1 "Implementing local democracy and participatory assessment methods". COWAM 2. Cooperative research on the governance of radioactive waste management, 17 pp

Forti V, Baldé CP, Kuehr R, Bel G (2020) The global e-waste monitor 2020. Quantities, flows and the circular economy potential. United Nations University (UNU)/United Nations Institute for Training and Research (UNITAR)—co-hosted SCYCLE programme, International Telecommunication Union (ITU) and International Solid Waste Association (ISWA), Bonn/Geneva/Rotterdam, 120 pp

Freiburghaus D, Zimmermann W (1985) Wie wird Forschung politisch relevant? Erfahrungen in und mit den schweizerischen nationalen Forschungsprogrammen [How does research become politically relevant? Swiss national research programmes experience]. Paul Haupt, Bern

GNRP, Global Network Resiliency Platform (2022, web) Welcome to REG4COVID. Best practices to improve COVID-19 responses. #REG4COVID. https://reg4covid.itu.int/

Gross Stein J (2002) The cult of efficiency. Anansi Press, Toronto

Guba EG, Lincoln YS (1989) Fourth generation evaluation. Sage, Newbury Park

Helbling D (2009) Managing complexity in socio-economic systems. Eur Rev 17(2):423–438. https://doi.org/10.1017/S1062798709000775

HM Treasury (2020) Magenta book. Central Government guidance on evaluation. Information Policy Team, The National Archives, Kew, London. https://www.gov.uk/government/publications/the-magenta-book, https://www.gov.uk/official-documents

IAEA, International Atomic Energy Agency (2019) Application of multicriteria decision analysis methods to comparative evaluation of nuclear energy system options: final report of the INPRO collaborative project KIND. IAEA Nuclear Energy Series, No. NG-T-3.20. IAEA, Vienna. http://www.iaea.org/publications/index.html

IGDTP, Implementing geological disposal of radioactive waste technology platform (2022, web). https://igdtp.eu/

Interact (2022, web) Practical handbook for ongoing evaluation of territorial cooperation programmes. https://www.interact-eu.net/#events-library

ITU, International Telecommunication Union (2020) Covid-19 initiatives, partnerships and activities. 9 Sept 2020. https://www.itu.int/en/SiteAssets/COVID-19/ITU-COVID-19-activities.pdf

KIT (2020, web) Multi-criteria decision support for sustainability assessment of energy technologies. Research project of Karlsruhe Institute of Technology, KIT. https://www.itas.kit.edu/english/projects_baha20_toolent.php

Kovalenko T, Sornette D (2013) Dynamical diagnosis and solutions for resilient natural and social systems. GRF Davos Planet@Risk 1(1): 7–33

Krütli P, Stauffacher M, Flüeler T, Scholz RW (2010b) Functional-dynamic public participation in technological decision making: site selection processes of nuclear waste repositories. J Risk Res 13 (7):861–875

Kunzig R (2020) Is a world without trash possible? Nat Geogr Mag. March issue

Lai A, Hensley J, Krütli P, Stauffacher M (2016) (eds) Solid waste management in the Seychelles. USYS TdLab transdisciplinary case study 2016. ETH Zürich, USYS TdLab, p 12

Lay M, Papadopoulos I (2007) An exploration of fourth generation evaluation in practice. Evaluation 13(4):486–495. https://doi.org/10.1177/1356389007082135

MCDA, International Society on MCDM, multiple criteria decision making. http://www.mcdmsociety.org/content/publications

NEA, Nuclear Energy Agency (2022) Website (Homepage>Topics>radioactive waste management). https://www.oecd-nea.org/jcms/c_

12892/radioactive-waste-management. Societal aspects: https://www.oecd-nea.org/jcms/pl_29867/societal-aspects-of-radioactive-waste-management

OECD (2022, web) Better criteria for better evaluation. Revised evaluation criteria. Definitions and principles for use. OECD/DAC network on development evaluation. https://www.oecd.org/dac/evaluation/daccriteriaforevaluatingdevelopmentassistance.htm

Saarikoski H, Barton DN, Mustajoki J, Keune H, Gomez-Baggethun E, Langemeyer J (2016) Multi-criteria decision analysis (MCDA) in ecosystem service valuation. In: Potschin M, Jax K (eds) OpenNESS ecosystem services reference book. EC FP7 grant agreement no. 308428.

Sitex, Sustainable network for independent technical expertise (2022, web). http://sitexproject.eu/

Skeptical Science Team (2022, web) Global warming & climate change myths. Getting skeptical about global warming skepticism. https://skepticalscience.com/argument.php

Sornette D, Kröger W, Wheatley S (2019) New ways and needs for exploiting nuclear energy. Springer, Cham, Switzerland

Sportanddev.org (2022, web) What is monitoring and evaluation (M&E)? https://www.sportanddev.org/en/toolkit/monitoring-and-evaluation/what-monitoring-and-evaluation-me

Stauffacher M, Krütli P, Flüeler T, Scholz RW (2012) Learning from the transdisciplinary case study approach: a functional-dynamic approach to collaboration among diverse actors in applied energy settings (Chap. 11). In: Spreng D, Flüeler T, Goldblatt DL, Minsch J (eds) Tackling long-term global energy problems: the contribution of social science. Environment & Policy, vol 52. Springer, Dordrecht NL, pp 227–245

Treggiden K (2020) The resources we need are no longer in the ground, but in landfill. De zeen Magazine, 28 Oct 2020. https://www.dezeen.com/2020/10/28/circular-economy-katie-treggiden-wasted-when-trash-becomes-treasure/

User Participation (2022, web) Toolbox of smart participatory methods and tools. https://www.user-participation.eu/

von Winterfeldt D, Edwards W (1984) Patterns of conflict about risky technologies. Risk Anal 4(1):55–68

WHO (2020) Coronavirus disease (COVID-19) advice for the public: mythbusters. https://www.who.int/emergencies/diseases/novel-coronavirus-2019/advice-for-public/myth-busters

Yap JYL, Ho CC, Ting CY (2019) A systematic review of the applications of multi-criteria decision-making methods in site selection problems. Built Environ Project Asset Manage 9(4):548–563. https://doi.org/10.1108/BEPAM-05-2018-0078

Experience with Societal Dialogues

CBC (2020) Indigenous community votes down proposed nuclear waste bunker near Lake Huron. 1 Feb 2020. https://www.cbc.ca/news/canada/toronto/ont-nuclear-bunker-1.5448819

CNDP, Commission nationale du débat public (2022, web) Compte rendu et bilan du débat sur le PNGMDR [Taking stock of the national plan on radioactive material and waste management]. 25 Nov 2019. https://pngmdr.debatpublic.fr/

Evers A, Nowotny H (1989) Über den Umgang mit Unsicherheit. Die Entdeckung der Gestaltbarkeit von Gesellschaft [How to deal with uncertainty. The discovery of how to shape society]. Suhrkamp, Frankfurt aM, p 247

Feenstra CFJ, Mikunda T, Brunsting S (2010) What happened in Barendrecht? Case study on the planned onshore carbon dioxide storage in Barendrecht, The Netherlands. Energy research Centre of The Netherlands (ECN), Petten NL, 42 pp

InSOTEC (2014) Addressing the long-term management of high-level and long-lived nuclear wastes as a socio-technical problem: insights from InSOTEC. European Commission community research. Contract number: 269906. Deliverable (D 4.1). [Universiteit Antwerpen]. https://sites.google.com/a/insotec.eu/insotec/publications/final-report

Kasperson RE, Golding D, Tuler S (1992) Social distrust as a factor in siting hazardous facilities and communicating risks. Soc Issues 48(4):161–187. https://doi.org/10.1111/j.1540-4560.1992.tb01950.x

NWMO, Nuclear Waste Management Organization (2005) Choosing a way forward. The future management of Canada's used nuclear fuel. Final study. NWMO, Toronto

NWMO (2022, web) Areas no longer being studied/Study areas. https://www.nwmo.ca/en/Site-selection/Study-Areas/Areas-No-Longer-Being-Studied

SFOE, Swiss Federal Office of Energy (2020) Sectoral plan for deep geological repositories. https://www.bfe.admin.ch/bfe/en/home/supply/nuclear-energy/radioactive-waste/deep-geological-repositories-sectoral-plan.html

Sweden, Ministry of the Environment (2022) Government to permit final disposal of spent nuclear fuel at Forsmark. News. 27 January 2022. https://www.government.se/articles/2022/01/final-disposal-of-spent-nuclear-fuel/

Upham P, Roberts T (2011) Public perceptions of CCS in context: results of NearCO$_2$ focus groups in the UK, Belgium, The Netherlands, Germany, Spain and Poland. Energy Procedia 4:6338–6344

WNWR, World Nuclear Waste Report (2019) The world nuclear waste report 2019. Focus Europe. https://worldnuclearwastereport.org/

Wynne B (1980) Technology, risk, and participation: the social treatment of uncertainty. In: Conrad J (ed) Society, technology, and risk assessment. Academic Press, London, pp 83–107

Yellowhead Institute (2020) The lake is speaking to us: nuclear waste in Saugeen Ojibway Nation territory? https://yellowheadinstitute.org/2020/01/29/gganoonigonaa-zaagigan-the-lake-is-speaking-to-us-nuclear-waste-in-saugeen-ojibway-nation-territory/

Websites

https://www.iaea.org/ (International Atomic Energy Agency: radioactive waste)

https://www.iaea.org/publications

https://www.iaea.org/publications/14741/country-nuclear-power-profiles

http://www.oecd-nea.org/ (Nuclear Energy Agency, OECD: radioactive waste)

https://www.eea.europa.eu/themes/waste (European Environment Agency: conventional waste within European Union)

https://sdgs.un.org/partnerships (Global partnership on waste management under the UN sustainable development goals)

https://www.iea.org/fuels-and-technologies/carbon-capture-utilisation-and-storage (Internal Energy Agency: CCS/CCUS)

https://www.iea.org/programmes/clean-energy-transitions-programme

Sources

AP, Associated Press (2020) People take part in a 'We Do Not Consent' rally at Trafalgar square, organised by Stop New Normal, to protest against coronavirus restrictions, in London, Saturday, Sept. 26, 2020 (Fig. 7.3b)

Democritos (-370). In: Freeman K (1946) The Presocratic philosophers. A companion to Diels. Fragment 269. Basil Blackwell, Oxford. https://normanrentrop.de/en/courage-to-act/ (quote Sect. 7.7)

Fleck L (1935, 1979b) Genesis and development of a scientific fact. Trenn TJ, Merton RK (eds), transl. Bradley F, Trenn TJ. University of Chicago Press, Chicago, IL, pp 44, 51. (Orig.: Fleck L (1935, 1980)) Entstehung und Entwicklung einer wissenschaftlichen Tatsache. Einführung in die Lehre vom Denkstil und Denkkollektiv. In: Schäfer L, Schnelle T (eds) Suhrkamp Taschenbuch Wissenschaft, Frankfurt aM, pp 60, 70 (Box 7.5)

Flüeler T (2006b) Decision making for complex socio-technical systems. Robustness from lessons learned in long-term radioactive waste governance. Series Environment and Policy, vol 42. Springer, Dordrecht NL, p 261 (Box 7.1)

Geertz G (1973) Thick description. Toward an interpretive theory of culture. In: The interpretation of cultures: selected essays. Basic Books, New York, pp 3–30. http://hypergeertz.jku.at/geertztexts/thick_description.htm, from: http://xroads.virginia.edu/∼DRBR/geertz2.txt (Box 7.1)

von Goethe JW (1832) Faust. Der Tragödie zweyter Theil in fünf Acten. In: Deutsches Textarchiv. https://www.deutschestextarchiv.de/book/view/goethe_faust02_1832?p=348. Werke, vol 41. JG Cotta'sche Buchhandlung, Stuttgart/Tübingen, Germany. Faust II, Act 5, verse 11936f. https://quotepark.com/quotes/1859728-johann-wolfgang-von-goethe-who-strives-always-to-the-utmost-for-him-there-i/ (quote Sect. 7.7)

Great Barrington Declaration (2020, 4 Oct) As infectious disease epidemiologists and public health scientists we have grave concerns about the damaging physical and mental health impacts of the prevailing COVID-19 policies, and recommend an approach we call focused protection. https://gbdeclaration.org/ (footnote 2)

IAP2, International Association for Public Participation (2022, web) Pillars for P2 brochure. IAP2 international headquarters, Denver. © IAP2 international federation 2018. All rights reserved. 20181112_v1. https://www.iap2.org/page/pillars (Fig. 7.2)

John Snow Memorandum (2020, 15 Oct) https://www.johnsnowmemo.com/ (Scientific consensus on the COVID-19 pandemic: we need to act now. Lancet 296:e71–e72. https://doi.org/10.1016/S0140-6736(20)32153-X (footnote 2)

Lochard J (2002) Concluding remarks. In: NEA (ed) Better integration of radiation protection in modern society. [2nd] Workshop proceedings. Villigen, Switzerland, 23–25 January 2001. OECD, Paris, pp 263–264; 263 (Box 7.1)

McDonnell J (2020) A very long line of voters wait to cast their ballots at the Fairfax County Government Center on first day of early voting in Fairfax, Virginia on Sept 18, 2020. Photo by John McDonnell/The Washington Post via Getty Images) (Fig. 7.3c2)

Noveck BS (2020) Democracy suffers when government statistics fail. Book review, 29 Sept 2020. Nature 586:27. https://www.nature.com/articles/d41586-020-02733-3 (Box 7.4)

Otway H (1987) Experts, risk communication, and democracy. Risk Analysis 7(2):125–129; 125 (Box 7.1)

Parker FL, Broshears RE, Pasztor J (1984) The disposal of high-level radioactive waste 1984. A comparative analysis of the state-of-the-art in selected countries. Volume I. NAK Rapport 11. Swedish National Board for Spent Nuclear Fuel. Beijer Institute, Royal Swedish Academy of Sciences, Stockholm, pp 116 (quote Sect. 7.7)

Popper K (1982) The open universe. An argument for indeterminism. From the "Postscript to the logic of scientific discovery". Routledge, London (quote Sect. 7.6)

Rittel H, Webber M (1973) Dilemmas in a general theory of planning. Policy Sci 4(2):155–169; 167, 169 (Box 7.1)

SFOE, Swiss Federal Office of Energy (2008) (Fig. 7.3c1)

TEPCO, Tokyo Electric Power Company (2011) Press conference, May 2011 (Fig. 7.3a)

WHO, Tedros AG (2020b) WHO Director-General's opening remarks at the [73rd] World Health Assembly, 18 May 2020. https://www.who.int/director-general/speeches/detail/who-director-general-s-opening-remarks-at-the-world-health-assembly (reference Sect. 7.4)

Annex

Own contributions over time (from several perspectives: (a) through (k), see at end).

All weblinks accessed 27 January 2023.

1.
Technical and institutional analyses
Case studies
Radioactive waste (SES 1987a, 1997, KSA 1998, KFW 2001, 2002d, KFW 2002ab, Ahlström 2003, 2003bd, 2004bc, 2005d, KSA 2005, 2006b, 2007a Junker 2008, 2009b, 2010b, AdK 2010, AG SiKa/KES 2011, 2012abd, RCJU 2012ff, 2013, 2014a, Seidl 2017, AG SiKa/KES 2017, AdK 2017, 2019, AG SiKa/KES 2019/2020, RCJU 2021ff, Seidl 2021
Conventional waste 1998b, 2000a, 2003d, 2012b, 2013, 2014a, Seidl 2017/2021
CCS 2012d, 2014a
Comparative studies 1998b, 2000a, 2003d, 2012d, 2013, 2014a, Seidl 2017/2021, 2021, 2022c, 2023

2.
Integral robustness … resilience, multiperspectiveness
1998ab, 1999, 2001acd, 2002acdf, 2003cd, 2004ab, 2004bc, 2005bc, 2005ef, 2006abc, 2007abcde, COWAM 2 2007abc, 2007, 2009ab, AdK 2010, 2012abc, Goldblatt 2012, Minsch 2012ab, 2013, 2014abcde, 2015, BABS 2015, 2016, Seidl 2017, AdK 2017, 2019, Seidl 2021

3.
Procedural and agent aspects
SES 1987ab, 1991, 1993, 1994, 1997, 1998cd, 2000bc, 2001b, IAEA 2001, 2002bde, KSA 2002, COWAM 2003, 2003ad, IAEA 2003, 2004ab, Hériard Dubreuil 2004, 2005aefgh, Stauffacher 2005, 2006abd, Krütli 2006, 2007abcde, COWAM 2 2007abc, Stauffacher 2008, Hériard Dubreuil 2008, Krütli 2009, Schori 2009, 2010abc, AdK 2010, 2012abc, Minsch 2012ab, Stauffacher 2012, 2013, 2014d, 2015, 2016, Seidl 2017, AdK 2017, 2019, Seidl 2021, 2021, 2022abcd, 2023

4.
Literacy in long-term governance of controversial socio-technical environmental issues. A guide for Strategic Monitoring of radioactive waste, conventional toxic waste, and carbon storage

---

Flüeler T (2023) Herausforderungen des Verfahrens und Lösungsansätze [Challenges to the procedure and ways of solving them]. In: Müller M. C. M. (ed) Die Suche nach einem Lager für hochradioaktive Abfälle. In der Schweiz und in Deutschland [In search of a repository for high-level radioactive waste. Switzerland and Germany]. Reihe Loccumer Protokolle, Band 26-2022. Rehburg-Loccum, Germany (e).

Flüeler T (2022d) Nuclear and conventional toxic wastes compared. Systemic reflections on similar yet different waste treatments. 1st TRANSENS Summer School: Transdisciplinary research for nuclear waste disposal: science meets society. 19-28 Aug, 2022, Bad Honnef (D). https://www.transens.de/en (e).

Flüeler T (2022c) Lessons from the Swiss approach – and others – to radwaste disposal. In search of sites for repositories for nuclear waste. 1st TRANSENS Summer School: Transdisciplinary research for nuclear waste disposal: science meets society. 19-28 Aug, 2022, Bad Honnef (D). https://www.transens.de/en (e).

RCJU, République et Canton du Jura, Département de l'environnement (2022) Projet Mont-Terri (MTP) et Centre de visiteurs (CV): Autorisation pour la phase 28. 1er juillet 2022. Département de l'environnement, Delémont, Switzerland (co-author) (k).

Flüeler T (2022b) Faden nicht abreissen lassen! Forum Endlagersuche als Fachöffentlichkeit und Bindeglied zu den künftigen Regionalkonferenzen [Handing over the torch! The Forum on Repository Search as a technical public and missing link to the future Regional Conferences]. 1. Forum Endlagersuche, May 20/21. Mainz, Germany. https://www.endlagersuche-infoplattform.de/ (e).

Flüeler T (2022a) Literacy in dealing with long-term controversial sociotechnical environmental issues. The cases of nuclear, special and carbon waste. Institute for Environmental Decisions Seminar, 10 May 2022. ETH, Zurich (e).

Flüeler T (2021) Öffentlichkeitsbeteiligung in der Beteiligungslücke nach Schritt 1 – aber sicher! Ein Blick von aussen auf und Empfehlungen für Schritt 2 der Phase 1 des Standortauswahlverfahrens für Endlager für hochradioaktive Abfälle [Public involvement in the participatory void after Step 1 – certainly and safely! An external view and some recommendations on Step 2 in Phase 1 of the (ongoing German) site-selection procedure for repositories of high-level radioactive waste]. Mandated by the National Citizens' Oversight Committee NBG, Berlin. https://www.nationales-begleitgremium.de/DE/WasWirMachen/Publikationen/publikationen_node.html (e).

RCJU, République et Canton du Jura, Département de l'environnement (2021) Projet Mont-Terri (MTP) et Centre de visiteurs (CV): Autorisation pour la phase 27. 1er juillet 2021. Département de l'environnement, Delémont, Switzerland, 8 pp (8 annexes) (co-author) (k).

Seidl R, Flüeler T, Krütli P (2021) Sharp discrepancies between nuclear and conventional toxic waste: technical analysis and public perception. Journal of Hazardous Material 414:125422. https://doi.org/10.1016/j.jhazmat.2021.125422 (g).

RCJU, République et Canton du Jura, Département de l'environnement (2020) Projet Mont-Terri (MTP) et Centre de visiteurs (CV): Autorisation pour la phase 26. 1er juillet 2020. Département de l'environnement, Delémont, Switzerland, 8 pp (8 annexes) (co-author) (k).

AG SiKa/KES, Arbeitsgruppe Sicherheit Kantone (AG SiKa)/Kantonale Expertengruppe Sicherheit (KES) [Cantons' Working Group on Safety/Cantonal Expert Group on Safety] (2020) ENSI-Richtlinie G03, Entwurf zur externen Anhörung. Stellungnahme [Nuclear regulator's guideline G03 on final radioactive waste disposal, draft for external consultation]. AWEL, Zürich, 11 pp (lead author, chair). https://www.zh.ch/de/umwelt-tiere/abfall-rohstoffe/radioaktive-abfaelle-tiefenlager.html (h).

AG SiKa/KES, Arbeitsgruppe Sicherheit Kantone (AG SiKa)/Kantonale Expertengruppe Sicherheit (KES) [Cantons' Working Group on Safety/Cantonal Expert Group on Safety] (2019) Entsorgungsprogramm 2016. Kommentar [Commentary on the Disposal Programme 2016]. AWEL, Zürich, 13 pp (lead author, chair) (h).

Flüeler T (2019) Strategic Monitoring—a proposal for the institutional surveillance of complex and long-term disposal programmes. In: MODERN 2020 Consortium (ed) Development and Demonstration of monitoring strategies and technologies for geological disposal. Contract No. 622177. Deliverable no 6.3. Final conference proceedings. 2nd International Conference on Monitoring in geological disposal of radioactive waste: strategies, technologies, decision making and public involvement. Paris, 9–11 Apr 2019. Project in the Euratom research and training programme 2014–2018 under grant agreement No 662177, pp 291–298. http://www.modern2020.eu/deliverables.html (g).

AWEL (2017) Strahlungsrisiken im Kanton Zürich. Auslegeordnung, Handlungsbedarf und Empfehlungen [Radiation risks in the Canton of Zurich. State of affairs, need for actions and recommendations]. AWEL, Zürich, pp 76 (co-author) (h).

AdK, Ausschuss der Kantone [Committee of the Cantons] (2017). Sachplan geologische Tiefenlager. Stellungnahme zu Etappe 2 [Sectoral plan geological repository. Statement concerning Phase 2]. AWEL, Zürich, pp 34 (co-author). https://www.zh.ch/de/umwelt-tiere/abfall-rohstoffe/radioaktive-abfaelle-tiefenlager.html (h).

AG SiKa/KES, Arbeitsgruppe Sicherheit Kantone (AG SiKa)/Kantonale Expertengruppe Sicherheit (KES) [Cantons' working group on safety/Cantonal expert group on safety] (2017) Sachplan geologische Tiefenlager (SGT). Etappe 2. Fachbericht zu Etappe 2 [Sectoral plan geological repository. Phase 2. Technical report on Phase 2]. AWEL, Zürich, pp 41 (lead author) (h).

Seidl R, Flüeler T, Krütli P, Moser C, Stauffacher M (2017) Radioaktive Abfälle und Sonderabfall im Vergleich [A comparison of radioactive and conventional toxic waste]. USYS TdLab. ETHZ, Zürich.

Flüeler T (2016) On the Final Report of the German Commission on Nuclear Waste Disposal. Reflections by an external observer. 2nd DAEF Conference on Key Topics in Deep Geological Disposal. Presentation, slide #7. Cologne, 26–28 Sep 2016. KIT, Köln, Germany (c, e, g).

BABS, Bundesamt für Bevölkerungsschutz [Federal Office for Civil Protection] (2015) Notfallschutzkonzept bei einem KKW-Unfall in der Schweiz [Emergency response concept in case of an accident at a nuclear power plant in Switzerland]. Ident-Nr./Vers. 10,013,284,897/01. BABS, Bern, pp 52 (co-author). https://www.admin.ch/gov/fr/accueil/documentation/communiques.msg-id-57954.html (j).

Flüeler T (2015) Inclusive assessment in a site-selection process—Approach, experience, reflections and some lessons beyond boundaries. In: Fanghänel S (ed) Deutsche Arbeitsgemeinschaft für Endlagerforschung (DAEF). Key topics in deep geological disposal. Conference report. Cologne, 24–26 Oct 2014. Karlsruher Scientific Reports, vol 7696. Karlsruher Institut für Technologie, Karlsruhe, Germany, pp 53–58. https://publikationen.bibliothek.kit.edu/1000047121 (g, h, i).

Flüeler T (2014e) Standortauswahl für geologische Tiefenlager als Entscheidungsfindung in komplexen soziotechnischen Systemen: Lehren aus internationalen Erfahrungen und Versuch einer ganzheitlichen Herangehensweise [Site selection for geological repositories as decision making in complex sociotechnical systems: lessons learnt from international experience and attempt to an integral approach]. Anhörung der Endlagerkommission, Paul-Loebe-Haus, Berlin, 5.12.2014.

Flüeler T (2014d) Inclusive assessment in a site-selection process—Approach, experience, reflections and some lessons beyond boundaries. DAEF Conference on Key Topics in Deep Geological Disposal. Cologne, 24–26 Oct 2014 (presentation). https://publikationen.bibliothek.kit.edu/1000047121 (g, h, i).

Flüeler T (2014c) Sachplan geologische Tiefenlager—aktuelle Einschätzung aus kantonaler Sicht [Sectoral plan on the site selection for geological repositories—current appraisal from a cantonal perspective]. Presentation at Mont-Terri Underground Laboratory visit by the Ministry of Environment of Lower Saxony. St-Ursanne, 2 Sept 2014 (slides) (h, i).

Flüeler T (2014b) Extended reviewing or the role of potential siting cantons in the ongoing Swiss site selection procedure ("Sectoral Plan"). In: NEA (ed) The safety case for deep geological disposal of radioactive waste: 2013 state

of the art. Symposium proceedings. Nuclear Energy Agency, Paris, 7–9 October 2013. NEA/RWM/R(2013)9. OECD, Paris, pp 405–412. https://www.oecd-nea.org/jcms/pl_19432/the-safety-case-for-deep-geological-disposal-of-radioactive-waste-2013-state-of-the-art?details=true (g, h, i).

Flüeler T (2014a) Radioactive and conventional toxic waste compared—an integrated approach, useful for an appraisal of carbon capture and storage (CCS). In: NEA (ed) The safety case for deep geological disposal of radioactive waste: 2013 state of the art. Symposium proceedings. Nuclear Energy Agency, Paris, 7–9 Oct 2013. NEA/RWM/R (2013)9. OECD, Paris, pp 351–361. https://www.oecd-nea.org/jcms/pl_19432/the-safety-case-for-deep-geological-disposal-of-radioactive-waste-2013-state-of-the-art?details=true, https://inis.iaea.org/collection/NCLCollectionStore/_Public/46/027/46027347.pdf (g).

Flüeler T (2013) Modul 3: Nuklearabfall im Vergleich mit Sonderabfall—Umgang und Wahrnehmung. Systemstudie, Teil I [Nuclear waste compared with conventional toxic special waste—handling, management, governance and perception. Systems study, Part I]. ETH-Swissnuclear-Kooperationsprojekt "Wege in eine Allianz der Verantwortung" [Joint project between ETH Zurich and Swissnuclear "Paths towards an alliance of responsibility"]. Monograph. ETH, Zürich (c, g).

RCJU, République et Canton du Jura, Département de l'environnement (2012) Projet Mont-Terri (MTP): Autorisation pour la phase 18. Département de l'environnement, Delémont, Switzerland (co-author) (and following years) (k).

Minsch J, Flüeler T (2012b) Die Schweiz, ein immerwährendes Gespräch. Multiple Krise, Komplexität, Unsicherheit, Konflikte: Anforderungen an eine Transformation zu einer Nachhaltigen Entwicklung & die Rolle von Wissenschaft und Bildung [Switzerland, an ongoing discourse. Multiple crises, complexity, uncertainty, conflicts: requirements for a transformation to sustainable development & the role of science and education]. In: Stiftung Zukunftsrat (ed) Haushalten & Wirtschaften. Bausteine für eine zukunftsfähige Wirtschafts- und Geldordnung. Zürich/Chur, Rüegger, pp 23–27 (g).

Flüeler T (2012d) Technical fixes under surveillance—CCS and lessons learned from the governance of long-term radioactive waste management (Chap. 10). In: Spreng D, Flüeler T, Goldblatt DL, Minsch J (eds) Tackling long-term global energy problems: the contribution of social science. Environment & Policy, vol 52. Springer, Dordrecht NL, pp 191–226. https://doi.org/10.1007/978-94-007-2333-7_10 (g).

Stauffacher M, Krütli P, Flüeler T, Scholz RW (2012) Learning from the transdisciplinary case study approach: a functional-dynamic approach to collaboration among diverse actors in applied energy settings (Chap. 11). In: Spreng D,

Flüeler T, Goldblatt DL, Minsch J (eds) Tackling long-term global energy problems: the contribution of social science. Environment & Policy, vol 52. Springer, Dordrecht NL, pp 227–245 (g).

Minsch J, Flüeler T, Goldblatt DL, Spreng D (2012a) Lessons for problem-solving energy research in the social sciences (Chap. 14). In: Spreng D, Flüeler T, Goldblatt D, Minsch J (eds) Tackling long-term global energy problems: the contribution of social science. Environment & Policy, vol 52. Springer, Dordrecht NL, pp 273–319 (g).

Flüeler T, Goldblatt DL, Minsch J, Spreng D (2012c) Energy-related challenges (Chap. 2). In: Spreng D, Flüeler T, Goldblatt DL, Minsch J (eds) Tackling long-term global energy problems: the contribution of social science. Environment & Policy, vol 52. Springer, Dordrecht NL, pp 11–22 (g).

Goldblatt DL, Minsch J, Flüeler T, Spreng D (2012) Introduction (Chap. 1). In: Spreng D, Flüeler T, Goldblatt DL, Minsch J (eds) Tackling long-term global energy problems: the contribution of social science. Environment & Policy, vol 52. Springer, Dordrecht NL, pp 3–10 (g).

Flüeler T (2012b) Insights from a comparison of conventional toxic and radioactive waste: Objective and project longevity of programmes. In: NEA (ed) The preservation of records, knowledge and memory (RK&M) across generations: scoping the issue. Workshop proceedings. Issy-les-Moulineaux, France, 11–13 October 2011. Radioactive Waste Management NEA/RWM/R(2012)6, June 2012. OECD, Paris, pp 25, 68. https://www.oecd-nea.org/jcms/pl_39202/the-preservation-of-records (c, g).

Flüeler T (2012a) Reflections on reversibility and retrievability by an "intermediate" stakeholder. In: NEA, Nuclear Energy Agency (ed) Reversibility and retrievability in planning for geological disposal of radioactive waste. Proceedings of the "R&R" International Conference and Dialogue. Reims, France, 14–17 Dec 2010. OECD, Paris, pp 161–166. https://www.oecd-nea.org/upload/docs/application/pdf/2019-12/6993-proceedings-rr-reims.pdf (g, h, i).

AG SiKa/KES, Arbeitsgruppe Sicherheit Kantone (AG SiKa)/Kantonale Expertengruppe Sicherheit (KES) [Cantons' Working Group on Safety/Cantonal Expert Group on Safety] (2011) Sachplan geologische Tiefenlager (SGT) Etappe 2. Fachbericht zu den ergänzenden Untersuchungen im Hinblick auf die Einengung [Sectoral plan geological repository. Phase 2. Technical report on the additional investigations with regard to [siting region] reduction]. AWEL, Zürich, 12 pp (co-author) (h).

AdK, Ausschuss der Kantone [Committee of the Cantons] (2010). Sachplan geologische Tiefenlager. Stellungnahme zu Etappe 1 [Sectoral plan geological repository. Statement concerning Phase 1]. Baudirektion Kanton Zürich, Zürich, pp 56 (co-author) (h).

Krütli P, Flüeler T, Stauffacher M, Wiek A, Scholz RW (2010b) Technical safety vs. public involvement? A case study on the unrealized project for the disposal of nuclear waste at Wellenberg (Switzerland). J Integr Environ Sci 7(3): 229–244. https://doi.org/10.1080/1943815X.2010.506879 (g).

Krütli P, Stauffacher M, Flüeler T, Scholz RW (2010a) Functional-dynamic public participation in technological decision making: site selection processes of nuclear waste repositories. J Risk Res 13(7):861–875. https://doi.org/10.1080/13669871003703252 (g).

Flüeler T (2010b) Rückholbarkeit versus sicherer Einschluss [Retrievability versus safe closure]. Impulsreferat. In: Hocke P, Arens G (eds) Die Endlagerung hochradioaktiver Abfälle. Gesellschaftliche Erwartungen und Anforderungen an die Langzeitsicherheit, Tagungsdokumentation zum „Internationalen Endlagersymposium Berlin, 30.10. bis 01.11.2008". Karlsruhe, Berlin, Bonn, S. 150–154 (slides). See https://www.endlagerforschung.de/assets/daef_broschuere_okt_2014.pdf (g).

Flüeler T (2010a) Partizipation im Schweizer Sachplan geologische Tiefenlager – ein Zwischenstand [Participation in the Swiss Sectoral plan deep geological repositories—an in interim status report]. Forum Endlager-Dialog, Hannover, 27. Januar 2010 (c, g, h).

Schori S, Krütli P, Stauffacher M, Flüeler T, Scholz RW (2009) Stakeholder preferences with respect to decision processes in repository site selection for radioactive waste. Lessons learnt from a multiple case study in Sweden and Switzerland. Monograph. ETH Zurich, Zurich. https://ethz.ch/content/dam/ethz/special-interest/usys/tdlab/docs/csproducts/Band_2008.pdf (g).

Flüeler T (2009b) Vergleich internationaler Sicherheitsanforderungen mit deutschen Sicherheitsanforderungen an ein Endlager für wärmeentwickelnde radioaktive Abfälle [A comparison of international safety requirements with German safety requirements of a repository for heat-generating radioactive waste]. Im Auftrag des Bundesamts für Strahlenschutz (BfS). Nov 2002. 45pp (c, e).

Flüeler T, Krütli P, Stauffacher M, Wiek A, Scholz RW (2009a) Lessons from failures in the governance of complex and contested socio-technical issues: the Wellenberg case of a radioactive waste repository project in Switzerland. Nordic Environmental Social Sciences 2009 Conference. London, 10–12 June 2009 (g).

Krütli P, Stauffacher M, Flüeler T, Scholz RW (2009) Siting for nuclear waste repositories: Do fair processes enable the acceptance of the outcome? VALDOR 2009 Conference. Stockholm, 8–11 June 2009 (g).

Hériard Dubreuil G, Mays C, Espejo R, Flüeler T, Schneider T, Gadbois S, Paixà A (2008) COWAM 2. Cooperative research on the governance of radioactive waste management. Final synthesis report. European Commission.

Nuclear Science and Technology. Directorate-General for Research, Euratom. Paris: Mutadis. pp 64. https://cordis.europa.eu/project/id/508856 (g).

Stauffacher M, Flüeler T, Krütli P, Scholz RW (2008) Analytic and dynamic approach to collaboration: a transdisciplinary case study on sustainable landscape development in a Swiss Prealpine region. Syst Pract Action Res 21 (6):409–422. http://doc.rero.ch/record/320415 (g).

Junker B, Flüeler T, Stauffacher M, Scholz RW (2008) Description of the safety case for long-term disposal of radioactive waste—the iterative safety analysis approach as utilized in Switzerland. Technical paper as part of the 'Long-term dimension of radioactive waste disposal: the role of the time dimension for risk perception' project. Natural and Social Science Interface (NSSI). ETH Zurich, Zürich (g).

Stauffacher M, Krütli P, Kämpfen B, Flüeler T, Scholz RW (2008) Gesellschaft und radioaktive Abfälle: Hauptbotschaften und Schlussfolgerungen einer schweizweiten Befragung [Society and radioactive wastes: main messages and conclusions of a Swiss national survey]. In: Stauffacher M, Krütli P, Scholz RW (eds) Gesellschaft und radioaktive Abfälle. Ergebnisse einer schweizweiten Befragung. Zürich/Chur, Rüegger, pp 1–22. https://tdlab.usys.ethz.ch/publications/reports.html (g).

Flüeler T, Goldblatt D, Minsch J, Spreng D (2007) Meeting global energy challenges: towards an agenda for social-science research. Final Report for EFDA and BP. Energy Science Center, ETH Zurich (c, g).

COWAM 2 (ed) (2007c) National insights. WP 5 final report. Work Package 5 "National Insights". COWAM 2. Cooperative research on the governance of radioactive waste management. 156 pp (co-author, National Contact Person for Switzerland) (g).

COWAM 2 (ed) (2007b) Long term governance for radioactive waste management. Final report of COWAM 2 Work Package 4. Work Package 4 "Long-Term Governance of Radioactive Waste". COWAM 2. Cooperative research on the governance of radioactive waste management, 67 pp (co-author, expert) (g).

COWAM 2 (ed) (2007a) Roadmap for local committee construction. Better paths towards the governance of radioactive waste. Work Package 1 "Implementing Local Democracy and Participatory Assessment Methods". COWAM 2. Cooperative research on the governance of radioactive waste management, 40 pp (co-author, expert) (g).

Flüeler T (2007 h) What is "long term"? Definitions and implications. In: Schneider T, Schieber C, Lavelle S (eds) Long term governance for radioactive waste management. Annex of the Final Report of COWAM 2. Work Package 4, pp 53–56. Community Waste Management 2, EURATOM/FP7, FI6W-CT-508856. COWAM2-D4-12-A. https://cordis.europa.eu/project/id/508856/reporting (g).

Flüeler T, Krütli P, Stauffacher, M (2007g) Tools for local stakeholders in radioactive waste governance. Challenges and benefits of selected PTA techniques/short version. PTA-1. Work Package 1 "Implementing Local Democracy and Participatory Assessment Methods". COWAM 2. Cooperative research on the governance of radioactive waste management, 7pp (g).

Flüeler T, Krütli P, Stauffacher M (2007f) Tools for local stakeholders in radioactive waste governance/long version. Challenges and benefits of selected PTA techniques. PTA-1. Work Package 1 "Implementing Local Democracy and Participatory Assessment Methods". COWAM 2. Cooperative research on the governance of radioactive waste management, 17pp (g).

Flüeler T, Blowers A (2007e) Decision-making processes in radioactive waste governance. Insights and recommendations. Work Package 3 "Quality of decision-making processes". COWAM 2. Cooperative research on the governance of radioactive waste management. Feb 2007. 26pp. https://cordis.europa.eu/project/id/508856/reporting (g).

Flüeler T (2007d) Procesos de toma de decisiones en la gobernabilidad de los residuos radiactivos. Recomendaciones. Versión resumida. Work Package 3—Quality of Decision-making Processes. EU Concerted Action Project Community Waste Management COWAM 2, 12pp (g).

Flüeler T (2007c) Processus décisionels dans la gouvernance des déchets radioactifs. Recommendations. Version courte. EU Concerted Action Project Community Waste Management COWAM 2, 11pp (g).

Flüeler T (2007b) Decision-making processes in radioactive waste governance. Recommendations. Short version. Work package 3—quality of decision-making processes. EU Concerted Action Project Community Waste Management COWAM 2, 12pp (g).

Flüeler T (ed) (2007a) Decision-making processes in radioactive waste governance. Appendix: synopsis of national decision-making processes (Belgium, Czech Republic, France, Germany, Hungary, Netherlands, Romania, Slovenia, Spain, Sweden, Switzerland, United Kingdom). Work Package 3 "Quality of decision-making processes". COWAM 2. Cooperative research on the governance of radioactive waste management. Feb 2007. 72pp (g).

Krütli P, Stauffacher M, Flüeler T, Scholz RW (2006) Public involvement in repository site selection for nuclear waste: a dynamic view. In: Andersson K (ed.): VALDOR 2006. VALues in Decisions On Risk. Proceedings. Stockholm, May 14–18, 2006. SKI, Naturvardsverket, SGI, Formas, UK Nirex, OECD/NEA. Stockholm: Congrex Sweden AB, pp 96–105 (g).

Flüeler T (2006c) Panel discussion on "The nuclear industry about the principle of precaution: Can a decision be made about a geological repository?" EUROSAFE Forum

2006 on "Radioactive waste management: long term safety requirements and societal expectations". Paris, 13–14 Nov 2006 (c, e).

Flüeler T (2006b) Decision making for complex socio-technical systems. Robustness from lessons learned in long-term radioactive waste governance. Environment & Policy, vol 42. Springer, Dordrecht NL. https://doi.org/10.1007/1-4020-3529-2_9 (g).

Flüeler T (2006a) Von der Fachöffentlichkeit zum öffentlichen Diskurs. Schweizer Erfahrungen und Ansätze zu einem erweiterten Entscheidungsmodell [From the technical community to a public discourse. Swiss experience and approaches to an extended decision model]. In Hocke P, Grunwald A (eds) Wohin mit dem radioaktiven Abfall? Perspektiven für eine sozialwissenschaftliche Endlager-forschung [What route for radioactive waste? Social-science perspectives in repository research]. Gesellschaft–Technik–Umwelt. Neue Folge, vol 8. edition sigma, Berlin, pp 219–237 (ITAS Workshop "Zur Endlagerung radioaktiver Abfälle in Deutschland. Perspektiven für eine sozialwissenschaftliche Begleitforschung". Karlsruhe, 28/29 Oct 2005. Institute for Technology Assessment and Systems Analysis (ITAS), Research Center, Karlsruhe (g).

KSA, Swiss Federal Nuclear Safety Commission (2005) Stellungnahme zum Entsorgungsnachweis für abgebrannte Brennelemente, verglaste hochaktive sowie langlebige mittelaktive Abfälle (Projekt Opalinuston) [Statement on the demonstration of the disposal of spent nuclear fuel, vitrified high-level and long-lived medium-level waste (Project Opalinus Clay)]. KSA 23/170. Kommission für die Sicherheit von Kernanlagen, KSA, Villigen-PSI (lead author) (c, d).

Stauffacher M, Krütli P, Flüeler T (2005) Partizipative Verfahren der Technik-Bewertung: Erfahrungen und Herausforderungen unterschiedlicher Techniken [Participatory methods of technology assessment: experience and challenges of various techniques]. In: Zuberbühler A, Baggenstos M, Zoubek N, Janett A (eds) Strahlenschutz-Aspekte bei der Entsorgung radioaktiver Stoffe. 5. Gemeinsame Jahrestagung des Dt.-Schweiz. Fachverbandes für Strahlenschutz und des Österr. Verbandes für Strahlenschutz. Basel, 20.-23.9.2005. TÜV-Verlag, Köln, pp 492–499 (g).

Flüeler T, Krütli P, Stauffacher, M (2005h) Tools for local stakeholders in radioactive waste governance: challenges and benefits of selected participatory technology assessment techniques. Interim Report. Jan 2005. Contribution to the EU STREP Community Waste Management COWAM 2 (g).

Flüeler T (2005g) Des outils participatifs pour la gouvernance des déchets nucléaires : expérience et exigences de diverses méthodes. Les Entretiens européens. 3ème édition. "Gérer nos déchets nucléaires: choisir la sécurité et le développement durable". 25 novembre 2005, Reims (presentation) (g).

Flüeler T (2005f) Appraisal of the Canadian (NWMO's) attempt to trigger a societal discourse on radioactive waste options and governance in view of other countries' recent approaches. Description of proposed research project on behalf of the Canadian Studies Faculty Research Program. 22 Oct 2005 (accepted, grant received) (g).

Flüeler T (2005e) Proposed framework for decision-making processes. Paper for Working Group 3. EU Concerted Action project Community Waste Management COWAM 2. 5 July 2005, pp 7 (g).

Flüeler T (2005d) Kommentar zum Abschlussbericht von DBE u. a. (2005): "Untersuchung der Möglichkeiten und der sicherheitstechnischen Konsequenzen einer Option zur Rückholung eingelagerter Abfälle aus einem Endlager" [Comment on DBE, Investigation of possibilities and safety-relevant consequences of an option to retrieve disposed wastes from a repository]. Workshop "Sicherheitstechnische Einzelfragen der Endlagerung", 28./29.9.2005, Hannover. ETH Zürich, Zürich (c, e, g).

Flüeler T (2005c) Long-term knowledge generation and transfer in radioactive waste governance. A framework in response to the "Future as an Enlarged Tragedy of the Commons". In: Carrasquero JV, Welsch F, Oropeza A, Flüeler T, Callaos N (eds) Proceedings PISTA 2005. 3rd International Conference on Politics and Information Systems: Technologies and Applications. Orlando, Florida, 14–17 July 2005. International Institute of Informatics and Systemics, IIIS Copyright Manager, Orlando, FA, pp 20–25 (g).

Flüeler T (2005b) Long-term knowledge generation and transfer in radioactive waste governance. Setting the scene. In: Carrasquero JV, Welsch F, Oropeza A, Flüeler T, Callaos N (eds) Proceedings PISTA 2005. 3rd International Conference on Politics and Information Systems: Technologies and Applications. Orlando, Florida, 14–17 July 2005. International Institute of Informatics and Systemics, IIIS Copyright Manager, Orlando, FA, pp 1–3 (g).

Flüeler T (2005a) "Akzeptanz durch Partizipation?" [Acceptance via participation?] Bern, 29. Juni 2005. Tagungsbericht für "Technikfolgenabschätzung. Theorie und Praxis", Zeitschrift des Instituts für Technikfolgenabschätzung und Systemanalyse (ITAS) im Forschungszentrum Karlsruhe. Nr. 3. Dez.:145–148 (c, g).

Hériard Dubreuil G, Gadbois S, Appel D, Åhagen H, Flüeler T, Kelly N, Le Bars Y, Mobb S, Neerdael B, Prêtre S, Schneider T, Vila d'Abadal M (2004) Local communities in nuclear waste management: the COWAM European Project. Euradwaste '04—Community policy and research initiatives. Sixth European Commission Conference on the Management and Disposal of Radioactive Waste. Luxemburg, Mar 29–Apr 1 (e, g).

Flüeler T (2004c) Stellungnahme zum Dokument "Grundsätze für die sichere Endlagerung. Die Sicherheitsphilosophie des Bundesamtes für Strahlenschutz [Statement regarding the document "Fundamentals for safe disposal. The safety philosophy of the Federal Office for Radiation Protection"]. BfS-01/04. Gutachten für das BfS [expert report for BfS]. ETH Zürich, Zürich (c, e).

Flüeler T, Scholz RW (2004b) Socio-technical knowledge for robust decision making in radioactive waste governance. Risk, Decision and Policy 9(2):129–159. https://doi.org/10.1080/14664530490464806 (g).

Flüeler T (2004a) Long-term radioactive waste management: challenges and approaches to regulatory decision making. In: Spitzer C, Schmocker U, Dang VN (eds) Probabilistic safety assessment and management 2004. PSAM 7–ESREL '04. Berlin, June 14–18, vol 5, Springer, London, pp 2591–2596, 2593. https://link.springer.com/chapter/10.1007%2F978-0-85729-410-4_415 (c, d, e).

IAEA (2003) Report of Technical Meeting—Waste Safety Standards—Stakeholder dialogue. 24–27 February 2003, International Atomic Energy Agency IAEA, Vienna (internal report, co-author) (c, e).

Flüeler T, Seiler H (2003d) Risk-based regulation of technical risks: lessons learnt from case studies in Switzerland. Journal of Risk Research 6(3):213–231. https://doi.org/10.1080/1366987032000088856 (c, g).

Flüeler T (2003c) Robust decision making in radioactive waste management is process-oriented. In: Andersson K (ed) VALDOR 2003. VALues in Decisions On Risk. Proceedings. Stockholm, June 9–13, 2003. SCK•CEN, SKI, SSI, NKS, OECD/NEA, UK Nirex, Stockholm, pp 79–87 (g).

Flüeler T (2003b) Die Einbettung der Arbeit des AkEnd in den internationalen Kontext. Kommentar aus der Sicht eines Beobachters [Integration of AkEnd into the international context. Comment by an observer]. In: Dally A (ed) Atommüll und sozialer Friede. Strategien der Standortsuche für nukleare Endlager. Loccumer Protokolle 05/03. Evangelische Akademie, Rehburg-Loccum, Germany, pp 121–147 (c, e, g).

Flüeler T (2003a) An NGO's perspective on stakeholder involvement. IAEA Technical Meeting—Waste Safety Standards—Stakeholder dialogue. Feb 24–27. IAEA, Vienna (presentation) (c, e).

Ahlström PE, Flüeler T, Leijon B, Ström A (2003) Some comments on AkEnd: Selection procedure for repository sites. Experts' report for ILK (International Committee on Nuclear Technology, Internationale Länderkommission Kerntechnik), July 2003. Stockholm/Zürich/Augsburg (c).

COWAM, Community Waste Management (2003) COWAM Network: Nuclear waste management from a local perspective. Reflections for a better governance. Final report. Nov 2003. Mutadis, Paris, 58 pp (co-author) (g).

KSA, Federal Nuclear Safety Commission (2002) Methodik der Aufsicht über Kernanlagen. Teil 1: Sicherheitsanforderungen und Überwachung. Empfehlungen an die HSK [Methodology of the supervision of nuclear facilities.

Part 1: safety requirements and surveillance. Recommendation to the Federal Safety Inspectorate]. Juni 2002. KSA, Würenlingen (co-author) (d).

KFW, Kantonale Fachgruppe Wellenberg (2002b) Abfallinventar SMA Wellenberg [Waste inventory of the Wellenberg low- and medium-level radioactive waste project]. Juni 2002. Bericht im Auftrag des Regierungsrates des Kantons Nidwalden, Stans (co-author) (f).

KFW, Kantonale Fachgruppe Wellenberg (2002a) Bericht zur Standortwahl Wellenberg [Report on the site selection of Wellenberg]. Januar 2002. Bericht im Auftrag des Regierungsrates des Kantons Nidwalden, Stans (co-author) (f).

Flüeler T (2002f) Long-term radioactive waste management: challenges and approaches to regulatory decision making. HSK/IAEA/NEA Workshop on Regulatory Decision-Making Processes. Grandhotel Giessbach, Brienz, Switzerland. 15–18 Oct 2002 (presentation) (c, d, e).

Flüeler T (2002e) COWAM: Lessons for the governance of radioactive waste management. Decision making in complex socio-technical systems. Directorate-General for Research, European Research 2002 Conference. Brussels, Belgium. Nov 11–13 (Designated exhibitor, 6th European Community Research Framework Programme 2003–2006). https://cordis.europa.eu/project/id/FIKW-CT-2000-20072 (g).

Flüeler T (2002d) Radioaktive Abfälle in der Schweiz. Muster der Entscheidungsfindung in komplexen soziotechnischen Systemen [Radioactive waste in Switzerland. Patterns of decision making in complex sociotechnical systems]. Doctoral dissertation, No 14645. ETH Zürich. dissertation.de, Berlin. https://www.research-collection.ethz.ch/handle/20.500.11850/146946 (g).

Flüeler T, Kastenberg B, Wildi W, Hufschmid P (2002c) Proposal for anticipatory radioactive waste research: Towards robust and long-term radioactive waste management and governance. In collaboration with the Swiss Federal Nuclear Safety Inspectorate, HSK. Submission in a call of the US National Research Council for anticipatory research. 2002-05-30, 13 pp (c, g).

Flüeler T (2002b) Abfall und Aufsicht im Kernenergiegesetz. Von der Wiederaufarbeitungsdebatte verdrängt [Waste and oversight in the Nuclear Energy Act. Suppressed by the discourse on reprocessing]. Neue Zürcher Zeitung, NZZ (80/2002), 8.4.2002:11 (c).

Flüeler T (2002a) Passive safety and controllability in radioactive waste management. A Swiss attempt on the borderline. Topical Day 2002: Ethics on Radioactive Waste. How radiant is our future? Belgian Nuclear Research Centre SCK•CEN. Mol (B), 22 Jan 2002 (presentation) (g).

IAEA (2001) Senior Consultants Report on Societal Issues. May 2001. International Atomic Energy Agency, IAEA, Vienna (internal report) (co-author) (e).

KFW, Kantonale Fachgruppe Wellenberg (2001) Bericht der KFW zum Konzessionsgesuch Sondierstollen

Wellenberg der GNW [KFW's report on the licensing application regarding the exploratory shaft Wellenberg by GNW]. April 2001. Bericht im Auftrag des Regierungsrates des Kantons Nidwalden, Stans (co-author) (f).

Flüeler T (2001d). Integrated robustness in radioactive waste management. In: Hériard Dubreuil G (ed) COWAM: Oskarshamn Seminar Report. EU Concerted Action project Community Waste Management COWAM. 1st Seminar. Oskarshamn (S), 19–22 Sep 2001, pp 89–92 (g).

Flüeler T (2001c) Robustness in radioactive waste management. A contribution to decision making in complex socio-technical systems. In: Zio E, Demichela M, Piccinini N. (eds.) Safety & Reliability. Towards a Safer World. Proceedings of the European Conference on Safety and Reliability. ESREL 2001. Torino (I), 16–20 Sep 2001, vol 1. Politecnico di Torino Torino (I), pp 317–325 (g).

Flüeler T (2001b) Lagerung radioaktiver Abfälle in der Schweiz: "Erfahrungen mit dem Bürgerdialog" oder der Umgang mit Dissens [The disposition of radioactive waste in Switzerland: "Experience with the citizens' dialogue" or how to handle dissent]. Vortrag vor dem deutschen Arbeitskreis Auswahlverfahren Endlagerstandorte AkEnd. Hauptabteilung für die Sicherheit der Kernanlagen HSK, Würenlingen, 13.3.2001 (presentation) (g).

Flüeler T (2001a) Options in radioactive waste management revisited: A framework for robust decision making. Risk Analysis 21(4):787–799. https://doi.org/10.1111/0272-4332.214150 (g).

Flüeler T (2000c) Mutual learning [in] radioactive waste management: Transcontinental gridlock–transdisciplinary solution? In: Scholz RW et al. (eds) Transdisciplinarity: Joint problem-solving among science, technology and society. Proceedings of the International Transdisciplinarity 2000 Conference, Zurich, 27 Feb-1 Mar 2000. Workbook II: Mutual learning sessions, vol 2. Haffmanns Sachbuch, Zürich, pp 304–307 (g).

Flüeler T. (2000b) (ed) Mutual learning session 15. Radioactive waste management: Transcontinental gridlock–transdisciplinary solution? A process document of the Mutual Learning Session 15. Working paper. ETH, Chair of Environmental Sciences: Natural and Social Science Interface NSSI, Zurich (g).

Flüeler T, van Dorp F (2000a) Risikobasiertes Recht. Fallstudie Abfälle [Risk-based regulation. Case study on waste]. Schweiz. Nationalfonds, Proj. Nr. 1113–52,163.97. Risk Based Regulation—ein taugliches Konzept für das Sicherheitsrecht? [Risk-based law. Case study on waste. Swiss National Science Foundation, Project No. 1113–52,163.97. Risk Based Regulation—A suitable concept for technical safety law?]. Stämpfli, Bern (c, g).

Flüeler T (1999) Pursuit of sustainability: guardianship or final disposal? Input paper for the Workshop on Disposition of High-Level Radioactive Waste. National Research Council/Board on Radioactive Waste Management. Irvine, CA, 4–5 Nov 1999 (unpublished) (g).

KSA (1998) Aktuelle Fragen zur Entsorgung radioaktiver Abfälle in der Schweiz—Position der KSA [Current issues on the disposal of radioactive waste in Switzerland—KSA's position]. KSA 21/124. KSA, Villigen-PSI, Switzerland. (co-author) (d).

Flüeler T (1998d) Entscheidungsprozesse in der Entsorgung von Energieabfällen: Eindrücke eines Beobachters [Decision processes in the disposal of energy waste: impressions by an observer]. Weiterbildungskurs ETH Zürich/Forum VERA "Entscheidungsprozesse in der Entsorgung von Energieabfällen", Fürigen, 10.-12.9.1998 (presentation) (c).

Flüeler T (1998c) Vier Jahrzehnte Umgang mit radioaktiven Abfällen in der Schweiz—Perspektiven an Hand der Risikowahrnehmungsforschung [Four decades of radioactive waste management in Switzerland—Perspectives supported by risk perception research]. Unveröffentlichter Bericht. Bundesamt für Energie, Bern (Unpublished) (g).

Flüeler T (1998b) Decision anomalies and institutional learning in radioactive waste management. In: ANS (ed) Conference proceedings. 8th International Conference on High-Level Radioactive Waste Management, Las Vegas. American Nuclear Society, La Grange Park, IL, pp 796–799 (g).

Flüeler T (1998a) "Hüten oder endlagern?"—Ist das die Frage? ["Guarding or final disposal?" Is this the question?] Referat an der Veranstaltung zur "Problematik der radioaktiven Abfälle: Hüten? Endlagern? Kontrollieren?", 2.3.98. Parlamentarische Gruppe für Bildung, Wissenschaft, Forschung und Technologie, Bern (presentation) (c).

KSA, Swiss Federal Nuclear Safety Commission (1997) Safety culture in a nuclear installation. Reflections on its assessment and promotion. Feb 2002. KSA 7/75E. KSA, Würenlingen/Villigen-PSI, Switzerland, 61 pp (co-author) (c, d).

Flüeler T, Küppers C, Sailer M (1997) Die Wiederaufarbeitung von abgebrannten Brennelementen aus schweizerischen Atomkraftwerken. "Recycling" von atomarem Material aus der Schweiz im Ausland. Analysen der Konsequenzen für Umwelt und Energiepolitik [Reprocessing of spent nuclear fuel from Swiss nuclear power plants. "Recycling" of Swiss nuclear material abroad. Consequence analysis for the environment and energy policy]. CAN Anti Atom Koalition, Zürich (c).

Flüeler T (1994) Radioaktive Abfälle: Forderungen an die Langzeitlagerung [Radioactive wastes: requirements on long-term disposition]. Energie + Umwelt 1/94:8–11. SES, Zürich (a).

Flüeler T (1993) Wege aus dem Atommüll-Labyrinth? [Ways out of the atomic trash maze?] Energie + Umwelt 4/93:4–5 (a).

Flüeler T (1991) Atommüll—von der Sorge um die Nachsorge des Energiefriedens [Atomic trash—worries on the follow-up care of the "energy peace»]. Energie + Umwelt 2/91:4–5. SES, Zürich (b).

SES, Schweizerische Energie-Stiftung (1987b) NAGRA —Wie immer ohne Gewähr [As always: without any guarantee]. Energie + Umwelt [Journal Energy & Environment] 3/87. SES, Zürich, 24 pp (editor-in-chief). https://www.energiestiftung.ch/magazin-energie-und-umwelt.html (a).

SES, Schweizerische Energie-Stiftung (1987a) Projekt "Gewähr": Stellungnahme zu den Experten- und Kommissionsberichten des Bundes. Brief an Bundesamt für Energiewirtschaft [Project "Guarantee": statement on the Federal experts' and committees' reports. Letter to the Federal Office of Energy]. 26.2.1987. SES [Swiss Energy Foundation], Zürich (co-author) (a, c).

Own Perspectives Learned Over Time
(contributions above categorised by perspective below "a" through "k", in case of co-authorship first author/body listed).

- Science journalism (free lancer, editor, 1983–1986);
- Environmental NGOs (director of Swiss Energy Foundation, 1986–1990) (a);
- Conflict resolution issues and techniques: mandate by NGOs in Federal mediation attempt "conflict-solving group radioactive waste" (1991–1992) (b);
- Consultancy (independent advisor, from 1991) (c);

- Scientific advice to Government (Swiss Federal Nuclear Safety Commission, chair of Standing committee on radiation protection and waste disposal, 1992–2004, 2001–2004) (d)
- Expertise to, i.a., technical support organisations: IAEA, NEA, German Federal Office for Radiation Protection (BfS), German National Citizens' Oversight Committee NBG, Slovak Center for Nuclear Safety, etc. (from 2001) (e);
- Advice to Cantonal Government (Cantonal Technical Group Wellenberg, 2000–2002) (f);
- Academia: research and teaching (ETH Zurich, from 1996) (g);
- Public administration, civil service (Directorate of Public Works of the Canton of Zurich, Nuclear Technology Unit Head, 2009 to October 2022) (h);
- Stakeholding and technical expertise (partial state, i.e., canton) in the current Swiss site-selection process (2009 to October 2022) (i);
- Support for decision making regarding emergency and large-scale events and situations: Federal Civil Protection Crisis Management Board (member on behalf of the Conference of the Cantonal Energy Ministers, 2010 to October 2022) (j);
- Research oversight over research laboratory (Commission de suivi de la République et Canton du Jura, supervision of URL and advice to Government regarding the Underground Research Lab Mont-Terri, Switzerland, from 2012) (k).

Glossary, Indexed

It is intended to also give colloquial definitions so that non-experts can more easily relate to the respective issues. Always refer to the specified definitions in the main text and the indexed passages (in brackets). The legend for sources is at the end.

Acceptance, accepted (Sects. 2.7, 2.8, Figs. 3.1, 3.5)
General agreement that something is satisfactory or right, or that someone should be included in a group^
Generally agreed to be satisfactory or right^

Acceptability, acceptable
Satisfactory and able to be agreed to or approved of^

Accountability, accountable (Sects. 6.1.2, 6.2)
Completely responsible for what they do and must be able to give a satisfactory reason for it^

Actor (here also: agent, player, stakeholder)
Person who acts for or represents another^

Adaptive
Having an ability to change to suit different conditions^

Adaptiveness, adaptability, adaptivity (Box 5.2)
Ability or willingness to change in order to suit different conditions^

Agent (here also: actor, player, stakeholder)
Person who acts for or represents another^

Anthropocentrism, anthropocentric (Table 3.1)
Belief that humans are the most important entity in the universe; many proponents argue that a sound long-term view acknowledges that the global environment must be made continually suitable for humans and that the real issue is shallow anthropocentrism.°°°

Aquifer
Special underground rock layers that hold groundwater, which are often an important source of water for public water supply, agriculture and industry.°
Subsurface rock or sediment unit that is porous and permeable. To be an aquifer, it must have these traits to a high enough degree that it stores and transmits useful quantities of water***

Aquitard
Opposite of an aquifer. An aquiclude or aquitard is a subsurface rock, soil or sediment unit that does not yield useful quantities of water***

Archive (Box 7.2)
Collection of historical records relating to a place, organisation or family^

Assessment (Sect. 2.3, Fig. 3.4)
The process, and the result, of analysing systematically and evaluating the hazards associated with sources and practices, and associated protection and safety measures*
Act of judging or deciding the amount, value, quality, or importance of something, or the judgement or decision that is made^

Asymmetry (Sects. 2.7, 3.1, 5.7)
State of two halves, sides or parts that are not exactly the same in shape or size^

Intergenerational There is a "tyranny of the present" in the sense that present generations tend to prefer short-term gains (for themselves) over the reduction of greater harms and costs for future generations. Descendants do not live yet, they cannot bargain, reciprocate or punish us for what we do today

T. Flüeler, *Governance of Radioactive Waste, Special Waste and Carbon Storage*, Springer Textbooks in Earth Sciences, Geography and Environment, https://doi.org/10.1007/978-3-031-03902-7

Intragenerational There also is a "tyranny of the many" over a few in that a society, having benefited from electricity from nuclear power, puts a burden on one of its regions by siting a repository for its nuclear waste in one location

ATA/ATW, alpha-toxic waste (Fig. 4.4)
Radioactive waste with a content of alpha-emitters exceeding a value of 20,000 Becquerels per gram of conditioned waste^^ (Swiss definition, consult entries below)

Authority (Box 5.3, Sects. 6.2.2, 6.3)
The moral or legal right or ability to control^
Governmental body (at the national, regional or local level) that is responsible for policies and interventions, including the development of standards and the provision of guidance, for maintaining or improving the targeted good (health, etc.), and that has the legal power of enforcing such policies and interventions*

Back end (Sect. 2.2)
The part of an object or place that is furthest from the front^
Denotes the last part of a material cycle with the disposal of non-usable (not recycled) substances (as opposed to the front end)

Barrier (Table 4.3, 4.5, Figs. 4.5, 4.7, 4.8, Sects. 4.5.1, 4.5.3, 5.1)
Physical obstruction that prevents or inhibits the movement of people, radionuclides or some other phenomenon (e.g., fire), or provides shielding against radiation*

Barrier (= obstacle) (Fig. 1.1)
Anything that prevents people from being together or understanding each other^

Biocentrism, biocentric (Table 3.1)
Ethical point of view that extends inherent value to all living things, recognising Earth's organisms as central in importance°°°

Caprock (Fig. 4.7, Tables 4.3 and 4.5)
Harder or more resistant rock type overlying a weaker or less resistant rock type°°° (serving as a barrier to prevent the release of harmful substances into the environment, the biosphere)

Capture (Sect. 6.3, Table 6.2)
Act of taking someone as a prisoner, or taking something into your possession, especially by force^
Regulatory capture Political entity, policymaker or regulatory agency is co-opted to serve the commercial, ideological or political interests of a minor constituency, such as a particular geographic area, industry, profession or ideological group; a special interest is prioritised over the general interests of the public, leading to a net loss for society°°°

Carbon dioxide (Box 4.1, Tables 4.4, 4.5)
Colourless, inert and non-poisonous gas, produced by all aerobic organisms°°°

Carbon sequestration (here: Tables 4.4, 4.5)
The process of removing additional carbon from the atmosphere and depositing it in other "reservoirs", principally through changes in land use. In practical terms, the carbon sequestration occurs mostly through the expansion of forests (UNEP 2007)**

Carbon cycle
Biogeochemical cycle by which carbon is exchanged among the biosphere, pedosphere, geosphere, hydrosphere and atmosphere of the Earth°°°

Causality principle ($\rightarrow$ polluter-pays principle)

Clean Development Mechanism, CDM (Sect. 2.4)
One of the flexible mechanisms defined in the Kyoto Protocol (IPCC 2007) that provides for emissions reduction projects which generate Certified Emission Reduction units, CERs which may be traded in emissions trading schemes°°°

Carbon Capture, and Storage, CCS (sequestration) (esp. Chap. 2, Tables 4.4, 4.5)
One of the geoengineering methods, i.e., to remove carbon/carbon dioxide from the Earth's atmosphere: process of capturing carbon dioxide, transporting it to a storage site and depositing it where it will not enter the atmosphere°°°, includes several carbon capture techniques, bioenergy CCS and direct air capture DACCS (Fig. 2.4)

Carbon Capture, Utilisation and Storage CCUS
Like CCS but including the utilisation of CO_2

Chain of obligation (Box 3.2)
Normative principle to link present to future generations

Climate
The usual condition of the temperature, humidity, atmospheric pressure, wind, rainfall and other meteorological elements in an area of the Earth's surface for a long time°°°

Climate change
Long-term changes in temperature, precipitation, wind and all other aspects of the Earth's climate. Often regarded as a result of human activity and fossil fuel consumption°
Change of climate, which is attributed directly or indirectly to human activity that alters the composition of the global atmosphere and which is in addition to natural climate variability observed over comparable time periods (UNFCCC)**

Climate change adaptation
Adjustments to natural or human systems in response to actual or expected climatic factors or their effects, including changes in rainfall and rising temperatures, which moderate harm or exploit beneficial opportunities°

Climate change mitigation
Action to reduce the impact of human activity on the climate system, primarily through reducing greenhouse gas emissions°

Closure (of a facility) (Fig. 3.2 and Sect. 4.7)

The completion of all operations at some time after the emplacement of spent fuel or radioactive waste in a disposal facility. This includes the final engineering or other work required to bring the facility to a condition that will be safe in the long term*

Closure (of a discussion) (Sects. 3.3, 4.5.1 and 5.1)

Process for ending a debate (= formal discussion) in a parliament so there can be a vote^

In science: "when a consensus emerges that the 'truth' has been winnowed from the various interpretations" (Bijker et al. 1987, 12)

CO_2 (Tables 4.4, 4.5)

Carbon dioxide

Co-decision (Sects. 3.2 and 3.3)

Following the "co-production" of knowledge, society takes a socially robust decision on the basis of a multitude of values and interests as well as technical and safety-targeted evidence

Complex (Box 1.1, Sects. 2.1, 3.1, 3.2, 5.1, 7.5)

Involving a lot of different but related parts^

Complex system: tightly coupled, aggravating learning, processes incompletely known

Complexity

High number of diverse and interrelated elements*

Compromise (Table 3.1)

Agreement in an argument in which the people involved reduce their demands or change their opinion in order to agree^. Some goals are given up (by all involved), concessions are made

Confidence (Sects. 4.5, 7.3, Fig. 5.3)

Quality of being certain of your abilities or of having trust in people, plans or the future^

"related to process dependability, based on evidence that can be provided through transparency" (NEA 2013, 9)

Confinement (Sects. 4.5, Table 4.2)

Closely related in meaning to containment, but confinement is typically used to refer to the safety function of preventing the "escape" of radioactive material, whereas containment refers to achieving that function*

Conflict (Sect. 3.3, Table 3.1)

Active disagreement, as between opposing opinions or needs^

Struggle and a clash of interest, opinion, or even principles°°°, different judgements on facts, different interests, targets, values, (unequal) distribution (of information, of power, etc.), in relations (interpersonal)

Consensus, consent (Sect. 3.3, Table 3.1)

Generally accepted opinion or decision among a group of people^, agreement to a decision (no unanimity needed), all interests are sufficiently considered

Consent (Table 3.1)

Permission or agreement^, does not require full agreement but the status that nobody has significant objections

Consequence analysis (Sect. 4.5.1)

Part of the → safety analysis in the nuclear waste community

Constraint (Sects. 1.1, 2.2, Table 3.1)

Something that controls what you do by keeping you within particular limits^

Constraint, technological (Sects. 1.1, 2.2, Table 3.1)

The fact that something (here waste) was produced by an upstream activity/technology (nuclear, conventional) and that cannot be made to disappear

Construct (Sects. 4.2, 7.3, Boxes 7.3, 7.5)

Idea or imaginary situation^

Social construct: exists not in objective reality but as a result of human interaction—it exists because humans agree that it exists (basic notion of constructivism, a theory of knowledge in sociology)°°°

Context (Box 1.3, Fig. 1.5, Sect. 5.6)

Situation within which something exists or happens, and that can help explain it^

Control (Box 5.1, Sect. 5.1, Fig. 6.1)

Typically implies not only checking or monitoring something, but also ensuring that corrective or enforcement measures are taken if the results of the checking or monitoring indicate such a need "can carry the meanings of both the function and the means of 'directing, regulating, or restraining'" (controller, function, means), "'control' will be a shared function between the human controller and the various barriers that carry out the built-in controls" (NEA 2014b)

Control, institutional (Box 5.3, Sects. 5.3, 6.2)

Control of a radioactive waste site by an authority or institution designated under the laws of a State. This control may be active (monitoring, surveillance, remedial work) or passive (land use control) and may be a factor in the design of a facility (e.g., a near-surface disposal facility)*

Controversy

A lot of disagreement or argument about something, usually because it affects or is important to many people^

Conventional

Here: non-ionising, non-radioactive (waste)

Co-production (of knowledge) (Sects. 3.2, 6.2.2)
Simultaneous production of knowledge and social order (Jasanoff 1995)

Cost-benefit analysis (Sect. 2.4, 7.5.3)
Process of comparing the costs involved in doing something to the advantage or profit that it may bring^
Systematic technical and economic evaluation of the positive effects (benefits) and negative effects (disbenefits, including monetary costs) of undertaking an action (decision aiding technique commonly used in the optimisation of protection and safety)*

Critical group (Sect. 4.5.1)
A group of members of the public which is reasonably homogeneous with respect to its exposure to a given radiation source and is typical of individuals receiving the highest effective dose or equivalent dose (as applicable) from the given source*

Critical thinking (Box 7.3)
"… involves … (1) an attitude of being disposed to consider in a thoughtful way the problems and subjects that come within the range of one's experiences, (2) knowledge of the methods of logical inquiry and reasoning, and (3) some skill in applying those methods" (Glaser 1941)

Culture (Box 6.3)
The way of life, especially the general customs and beliefs, of a particular group of people at a particular time^
"Rules for living and functioning in society" (Samovar et al. 2012, 11)

Cumulative e.g., impact (Box 6.3)
A number of developments in a locality or a continuous activity over time that together may have an increased impact on the environment, local community or economy*

Data (Sect. 4.4.1, 4.5.1, 4.5, Boxes 7.3, 7.4)
Information, esp. facts or numbers, collected to be examined and considered and used to help with making decisions^ (cf. Information, Knowledge (Management))

Decide-announce-defend, DAD (Sects. 3.4, 5.6, Fig. 6.4)
Traditional linear decision model with few agents involved (proponent, experts, authority) and minimal participation of others (public, NGOs, etc.)

Decision (Sect. 6.1.1)
Choice that you make about something after thinking about several possibilities^
Deciding means selecting alternatives and willing to implement one of them

Decision anomaly (Sect. 3.2)
Behaviour of one player which cannot be understood by other actors (e.g., regarding the principle of sustainability) (Fig. 3.4)

Decision making (Fig. 5.7, 6.1, 6.2, 6.5)
Process of making choices, esp. important choices^
Process of deciding, the judgement made, the choice taken and, ideally, the decision implemented

Decommissioning
All steps leading to the release of a nuclear facility, other than a disposal facility, from regulatory control*

Defence in depth (Sects. 4.5.3, 4.7)
Hierarchical deployment of different levels of diverse equipment and procedures (multiple barriers) to prevent the escalation of anticipated operational occurrences and to maintain the effectiveness of physical barriers placed between a (radiation) source or (radioactive/toxic) material and workers, members of the public or the environment, in operational states and, for some barriers, in accident conditions*
Implemented primarily through the combination of a number of consecutive and independent levels of protection that would have to fail before harmful effects could be caused to people or to the environment*

Deficit model (Sects. 2.7, 6.2.2)
Mental model following the conventional linear decision model of "Decide-announce-defend": Official stakeholders explain technical solutions to the stakeholders and the public (no real communication but plain information, lack of legitimacy by the absence of public participation)

Deliberation (Sect. 3.3, Box 5.3)
Process of thoughtfully weighing options, emphasises the use of logic and reason as opposed to power struggle, creativity or dialogue. Group decisions are generally made after deliberation through a vote or consensus of those involved°°°

Democracy (Box 7.4)
Belief in freedom and equality between people, or a system of government based on this belief, in which power is either held by elected representatives or directly by the people themselves^

Design (Figs. 3.1, 4.8, 6.2, 6.5, Sects. 4.3, 4.5.1, 4.5.2, 4.6, Box 6.1)
Process and result of developing a concept, detailed plans, supporting calculations and specifications for a facility and its parts*

Dialogue (Sects. 3.4, 7.2, 7.6)
Serious exchange of opinion, esp. among people or groups that disagree^
"characterises an approach of collaboration or partnership between the institutional actors and the affected communities essentially, involving public participation in the decision-making process and mutual learning" (NEA 2103, 13)

Dimension (Sects. 1.2, 3.1, 3.3, Figs. 3.1, 5.4, Box 5.3)
Part or feature or way of considering something^

Discipline (Box 1.3, Sect. 2.1)
Particular area of study, especially a subject studied at a college or university^

Disciplinary, cross-, multi-, pluri-, inter-, trans- (Box 1.3)
(Intra-)disciplinary working within a single discipline
Crossdisciplinary viewing of one discipline from the perspective of another disciplinary perspective
Multi- or pluridisciplinary various and different disciplines work together, each drawing on their disciplinary knowledge
Interdisciplinary different disciplines integrate knowledge and methods, aiming at a synthesis of approaches
Transdisciplinary science reaches beyond disciplinary (and science-driven) perspectives (out to society)

Discourse (Sects. 3.3, 6.1, Box 5.2, Table 6.1)
(From Latin *discursus,* "running to and fro") identifies and describes written and spoken communications. In semantics and discourse analysis, a discourse is a conceptual generalisation of conversation°°°

Disposal
Emplacement of waste in an appropriate facility without the intention of retrieval*
In some states, the term disposal is used to include discharges of effluents to the environment. In some states, the term disposal is used administratively in such a way as to include, for example, incineration of waste or the transfer of waste between operators*
Retrieval would require deliberate action to regain access to the waste; it does not mean that retrieval is not possible*

Disposal system
The system of properties of the site for a disposal facility, design of the disposal facility, physical structures and items, procedures for control, characteristics of waste and other elements that contribute in different ways and over different timescales to the fulfilment of safety functions for disposal*
Geological disposal facility facility for radioactive waste disposal located underground (usually several hundred metres or more below the surface) in a stable geological formation to provide long-term isolation of radionuclides from the biosphere*

Disposition
Consigning of, or arrangements for the consigning of, radioactive waste for some specified (interim or final) destination, for example, for the purpose of processing, disposal or storage* (umbrella term for storage and disposal)

Dissent (Table 3.1)
Strong difference of opinion on a particular subject, especially about an official suggestion or plan or a popular belief^

Dose (Figs. 4.2, 4.8, Tables 4.2, 4.5)
Measure of the energy deposited by radiation in a target*

Ecocentrism, ecocentric (Table 3.1)
Denoting a nature-centred, as opposed to human-centred (i.e., anthropocentric), system of values, recognising Earth's interactive living and non-living systems as central in importance°°°

Effect (Sects. 5.4, 5.5, Box 1.1, Tables 4.2, 4.5)
Result of a particular influence^
Antagonistic biologic response to exposure to multiple substances that is less than would be expected if the known effects of the individual substances were added together (Table 4.2, Sect. 7.5.1)
Additive overall consequence which is the result of two chemicals acting together and which is the simple sum of the effects of the chemicals acting independently (Table 4.2)
Synergistic when the combined effect of several forces operating is greater than the sum of the separate effects of the forces (Table 4.2)
(https://www.greenfacts.org/glossary)

Effectiveness (Box 5.3, Sects. 6.2.2, 7.5.5)
Ability to be successful and produce the intended results^, do the right thing, rather: get the thing done
"It is fundamentally the confusion between effectiveness and efficiency that stands between doing the right things and doing things right. *There is surely nothing quite so useless as doing with great efficiency what should not be done at all"* (Drucker 1963)

Efficiency (Chap. 2, Sects. 5.4, 7.5.5)
Ratio of output (work, e.g., of a machine, product, benefit) to input (e.g., energy, cost, effort), do the thing right
Highest: least amount of input to achieve the highest amount of output

Environment
Natural all living and non-living things occurring naturally°°° (Fig. 3.1, Sect. 4.5)
Social the culture that an individual lives in, and the people and institutions with whom they interact°°° (Box 6.1)

Equitable
Treating everyone fairly and in the same way^

Equity (Sects. 5.8, 6.1.2)
The situation in which everyone is treated fairly and equally^

Ethics (Figure 3.1, Box 3.2, Sects. 6.2.3, 7.2)
The study of what is morally right and wrong, or a set of beliefs about what is morally right and wrong^

Evaluation (Figs. 6.2, 6.5, Sects. 6.5, 7.5.5)
Judgement or calculation of the quality, importance, amount or value of something^
Assessment of completed projects, programmes or phases thereof (sportanddev.org 2022, HM Treasury 2020, 6)

Experience (L. Fleck in "About the Author", Boxes 1.3, 6.3, 7.3, Sects. 5.5, 6.1.3, 6.2.4, 7.7, Fig. 5.7, Boxes 6.3, 7.3, Table 6.2)
(Process of getting) knowledge or skill from doing, seeing or feeling things^

Expert (Box 7.1)
Person with a high level of knowledge or skill relating to a particular subject or activity^

Expert blocking (Box 6.2)
Using experts for one's own purposes (exclusively) and preventing others to take advantage of them

Failure (Fig. 1.2, Sect. 2.8, 5.1)
Loss of the ability of a structure, system or component to function within acceptance criteria*
Result of, for example, a hardware fault, a software fault, a system fault, an operator error or a maintenance error*

Failure culture (Sect. 6.4, Box 6.2)
Set of shared values, goals and practices that encourage learning through experimentation—key ingredient of innovation leadership (Kühl 2017)

Far field (Sect. 4.5.1, Fig. 6.3)
The geosphere outside a disposal facility, comprising the surrounding geological strata, subsoil and rock*

Fix, quick, technological (Sects. 2.1, 5.6, 6.2.3)
Situation causing trouble or problems^ (esp. American English)
Way of solving a problem, especially an easy or quick one^

Framing (Sects. 1.1, 6.1.2, Fig. 5.7)
Way of structuring or presenting a problem or an issue, involves explaining and describing the context of the problem, comprises a set of concepts and theoretical perspectives on how individuals, groups and societies organise, perceive and communicate about reality°°°
The framing effect is a cognitive bias where people decide on options based on whether the options are presented with positive or negative connotations, e.g., as a loss or as a gain°°°

GDP, gross domestic product (Sects. 2.5, 7.4)
Total monetary or market value of all the finished goods and services produced within a country's borders in a specific time period (https://www.investopedia.com/terms/g/gdp.asp)

Generation (here: Sects. 2.7, 3.1, 5.2, 5.6, Box 6.3)
All the people of about the same age within a society or within a particular family^

Geoengineering (Box 2.2)
Deliberate large-scale intervention in the Earth's natural systems to counteract climate change, wide range of proposed geoengineering techniques (Oxford Martin School, 2020)**

Goal (Box 1.1, Chap. 3, Sect. 4.4, 5.1, 5.5, 6.1, 6.2, 7.7, Box 6.1, Table 6.1)
A purpose, or something that you want to achieve^, e.g., the protection of present and future generations and environments

Governance (Sect. 5.5, Box 5.3, 6.2)
Way in which an organisation is managed at the highest level, and the systems for doing this^

Greenhouse effect/global warming (Box 4.1, Fig. 4.3)
The gradual heating of the Earth due to greenhouse gases, leading to climate change and rising sea levels. Renewable energy, energy-efficient buildings and sustainable travel are examples of ways to help avert the greenhouse effect°

Groupthink (Fig. 3.4)
Process in which bad decisions are made by a group because its members do not want to express opinions, suggest new ideas, etc., that others may disagree with^

Guardian (Sects. 3.4, 5.2, 5.3, 6.2.3, 7.1, Fig. 7.1)
Person who has the legal right and responsibility of taking care of someone who is not responsible for his or her own care; someone who protects something^

Half-life (Box 4.1, Sect. 5.2)
For a radionuclide, the time required for the activity to decrease, by a radioactive decay process, by half*

Hazard (Sects. 1.2, 4.2, Fig. 4.1, Tables 4.2, 4.5)
The potential for harm or other detriments, especially for risks; a factor or condition that might operate against safety*, something that is dangerous and likely to cause damage^

Hazardous waste (Sect. 1.1, Box 4.1, Tables 4.2, 4.3)
Wastes that have the potential to cause harm to human health or the environment.
Wastes that exhibit one or more hazardous characteristics, such as being flammable, oxidising, poisonous, infectious, corrosive or ecotoxic (Basel Convention)**

High-level radioactive waste, HLW (Box 4.1, Tables 4.1, 4.2, Fig. 4.4)

Spent fuel assemblies (SF) not destined for reprocessing, and vitrified fission product solutions from reprocessing of spent fuel^^ (consult entries below)

Host rock (Sect. 4.5.1, Tables 4.3, 4.5)
Geological formation(s) where wastes are disposed of and which should ensure sufficient isolation from living organisms

Impact (Sects. 2.3, 2.5, 2.7, 4.3, 6.2.1, 7.5.5, Figs. 4.2, 4.8, Table 4.2)
Powerful effect that something, especially something new, has on a situation or person^

Indeterminacy (Sects. 7.4, 7.5, 7.5.4)
The state of not being measured, counted or clearly known^, furthest away from certainty, further than ambiguity

Indicator (Sects. 4.5.1, 4.6, 7.5.1, 7.6, Tables 4.2, 4.5, Fig. 4.8, Box 5.2)
Performance indicator characteristic of a process that can be observed, measured or trended to infer or directly indicate the current and future performance of the process, with particular emphasis on satisfactory performance for safety*

Information (Sects. 2.7, 4.5.1, 5.4, 5.6, 6.1.1, 6.1.2, 6.4, 7.2, 7.3, 7.5.1, 7.5.3, Box 7.3)
Facts about a situation, person, event, etc.; facts or details about a person, company, product, etc.^ (cf. Data, Knowledge)

Information gap (Sect. 2.7)
→ Deficit model

Infrastructure (Sects. 2.3, 2.6, 5.4)
Basic services necessary for development to take place, for example, roads, electricity, sewerage, water, education and health facilities°

Innovation (Sects. 2.1, 2.8, 7.3, Fig. 2.2, Table 6.2, Box 7.4)
(The use of) a new idea or method^, focus on newness, improvement and spread, something original and more effective and, as a consequence, new, that "breaks into" the market or society, more apt to involve the practical implementation of an invention (i.e., new/improved ability) to make a meaningful impact in a market or society, and not all innovations require a new invention°°°

Institution (Box 5.3, Sects. 1.5, 2.1, 2.4, 5.2, 5.3, 5.5, 5.6, 5.8, 6.2, 7.1, 7.3, 7.7, Figs. 3.4, 3.5, 5.2, 6.1, 7.1, Tables 4.1, 4.4, 6.2)
"formal and informal rules and norms that organise social, political and economic relations" (North 1990)

Interpretative flexibility (Sect. 2.1)
Each technological artefact has different—sometimes contradictory—meanings and interpretations for various groups (Bijker et al. 1987). Over time, as technologies are developed, the interpretative and design flexibility collapse through closure mechanisms.°°° In the present context, e.g., spent fuel, it is a resource for supporters and waste for opponents of the nuclear industry. Likewise are nuclear power plants either facilities generating climate-friendly and cost-effective electricity or primary sources of harmful radiation contaminating humans and the environment

Isolation (Sects. 4.5.1, 4.6, 5.2, Fig. 3.1, Table 4.2)
The physical separation and retention of radioactive waste away from people and from the environment*
Isolation of radioactive waste with its associated hazards in a disposal facility involves the minimisation of the influence of factors that could reduce the integrity of the disposal facility; provision for very low mobility of most long-lived radionuclides to impede their migration from the disposal facility, and making access to the waste by people difficult without special technical capabilities*
Design features are intended to provide isolation (a confinement function) for several hundreds of years for short-lived waste and for at least several thousand years for intermediate-level waste and high-level waste. Isolation is an inherent feature of geological disposal*

Knowledge (Sect. 2.7, Chap. 3, Sects. 5.5, 5.6, 6.1.1, 6.1.3, 6.2, 6.4, 7.2, 7.3, 7.5.1, 7.5.3, 7.7, Boxes 1.3, 6.3, 7.2, 7.3, 7.5)
Understanding of or information about a subject that you get by experience or study, either known by one person or by people generally^
Term often used to refer to bodies of facts and principles accumulated by humankind over the course of time*
Explicit knowledge that is contained in, for example, documents, drawings, calculations, designs, databases, procedures and manuals*
Tacit knowledge that is held in a person's mind and has typically not been captured or transferred in any form (if it were, it would become explicit knowledge)*

Knowledge management
An integrated, systematic approach to identifying, managing and sharing an organisation's knowledge and enabling groups of people to create new knowledge collectively to help in achieving the organisation's objectives; helps an organisation gain insight and understanding from its own experience*
Knowledge is distinct from **information**: Data yield information and knowledge are gained by acquiring, understanding and interpreting information. Knowledge and information each consist of true statements, but knowledge serves a purpose: Knowledge confers a capacity for effective action*
The overall process is how to turn data, information and knowledge into cognisance and understanding

Landfill (Fig. 4.6, Sects. 4.4.1, 4.5.2, 5.6, 6.2.1, 7.5.6, Tables 4.1, 4.2, 4.3, 6.1)
Permanent disposal of waste into the ground, by the filling of man-made voids or similar features, or the construction of landforms above ground level (land-raising)°

Large technical systems, LTS (Box 1.4)
Mainly infrastructure systems with specific functions (e.g., electricity distribution), particular knowledge bases and norms, specialised professional groups. They are characterised by a high degree of technical and social integration.

Their interconnectedness may yield to more technical disruption and/or political interference. They should not be confounded with large-scale technical projects like the Manhattan project to construct an atomic bomb. Waste management and disposal are typical infrastructure systems, with complex technical components and organisational networks, subject to a high degree of regulation (by state laws and/or technical norms)

LD50/LD$_{50}$ (Table 4.2)
Test in toxicology: measure of the (median) lethal dose of a toxin, radiation or pathogen. The value of LD$_{50}$ for a substance is the dose required to kill half the members of a tested population after a specified test duration°°°

Learning (Sects. 5.6, 6.1.2, 6.2.4, 6.4, 7.1, 7.2, 7.3, 7.5.5, 7.7, Figs. 5.7, 6.4, Table 6.2, Box 7.1)
The activity of obtaining knowledge^
Double-loop entails the modification of goals or decision-making rules in the light of experience. The first loop uses the goals or decision-making rules, the second loop enables their modification, hence "double-loop". Double-loop learning recognises that the way a problem is defined and solved can be a source of the problem (Argyris 1991)

Learning, organisational (Sects. 5.6, 6.2.4, Table 6.2)
Process of recognising and correcting errors (Argyris 1982)

Legitimate, legitimacy (Sects. 3.1, 5.6, 6.2.2, 6.2.3, 6.2.4)
Allowed by law; reasonable and acceptable^

Literacy (Sects. 7.1, 7.2, 7.4, 7.7)
Knowledge of a particular subject, or a particular type of knowledge; basic skill or knowledge of a subject^

Lock-in (Sects. 2.6, 2.8, 6.3, Table 6.2)
Situation where a society/institution/group is caught, e.g., in technology or technologies, type of path dependency where decisions depend on prior decisions or past activities, foreclosing alternative ways (Question 2, Chap. 2)

Longevity (Sects. 4.3, 4.4.2, 4.6, 5.1, 5.6, 7.4, Fig. 5.4)
Living for a long time^, typical length of time°°°

Long-term (Sects. 1.2, 2.1, 2.3, 3.1, 4.3, 4.4.1, 4.5.1, 4.5.2, 4.5.3, 5.2, 5.3, 5.5, 5.6, Figs.3.2, 4.2, 5.4)
Cf. Section 5.2, Fig. 3.2, 5.4

Low- and intermediate-level radioactive waste, LILW (Box 4.1, Sect. 6.2.4, Tables 4.1, 4.3, 4.5, Figs. 4.3, 4.4)
Cf. Box 4.1. All other than HLW and ATW (above)

Maintenance (Sect. 4.4.1)
The organised activity, both administrative and technical, of keeping structures, systems and components in good operating condition, including both preventive and corrective (or repair) aspects*

Management (Sects. 2.1, 2.3, 4.4.1, 4.5.1, 4.5.3, 5.1, 5.2, 5.6, 6.2.3, 7.1, 7.7, Figs. 3.2, 4.5, 6.5, Table 4.5)
The control and organisation of something^
Integrated management system single coherent management system for facilities and activities in which all the component parts of an organisation (organisational structure, resources and organisational processes) are integrated to enable the organisation's objectives to be achieved*

Model (Fig. 3.5, 6.1, 6.2, Table 3.1, 6.2, Box 5.1, Sects. 3.2, 6.2.2)
Cf. Figure 3.5. Analytical or physical representation or quantification of a real system and the ways in which phenomena occur within that system, used to predict or assess the behaviour of the real system under specified (often hypothetical) conditions*
Conceptual set of qualitative assumptions used to describe a system (or part thereof). These assumptions would normally cover, as a minimum, the geometry and dimensionality of the system, initial and boundary conditions, time dependence and the nature of the relevant physical, chemical and biological processes and phenomena*

Monitoring (Sects. 2.5, 4.5, 5.1, 5.3, Chap. 6, Sect. 7.5.5, Fig. 6.3, Table 6.2)
Cf. Figure 6.3, Table 6.2. The measurement of an entity for reasons relating to the assessment or control of exposure or exposure due to hazardous substances, and the interpretation of the results. May be subdivided in different ways, according to location (e.g., source as opposed to environmental) or purpose (e.g., routine)*
"Checking progress against plans" (sportanddev.org 2022, HM Treasury 2020, 6)

Monitoring, strategic (Sects. 6.1.3, 6.5, 7.7, Table 6.2)
Governance of process, implementation or policy and institutional surveillance (adaptive management tool)

Multiperspectiveness (Sects. 2.1, 2.6, 2.8, 3.1, 3.2, 3.3)
Looking at, even integrating several ways of considering something

Multiple barriers
Two or more natural or engineered barriers used to isolate radioactive waste in, and to prevent or inhibit the migration of radionuclides from, a disposal facility*

Multiple-barrier concept
Cf. Figures 4.5, 4.8

Near field (Sect. 4.5.1, Fig. 4.8, 6.3)
The excavated area of a disposal facility near or in contact with the waste packages, including filling or sealing materials, and those parts of the host medium/rock whose characteristics have been or could be altered by the disposal facility or its contents*

Negative emissions
Removal of greenhouse gases, GHGs from the atmosphere by deliberate human activities, i.e., in addition to the removal that would occur via natural carbon cycle processes (IPCC, 2018)**

Net zero emissions (Sects. 2.3, 2.5, 4.5.3, 7.5.6)
Net zero emissions are achieved when anthropogenic emissions of greenhouse gases to the atmosphere are balanced by anthropogenic removals over a specified period. Where multiple greenhouse gases are involved, the quantification of net zero emissions depends on the climate metric chosen to compare emissions of different gases (such as global warming potential, global temperature change potential, and others, as well as the chosen time horizon) (IPCC 2018)**

Norm (Sects. 3.1, 3.4, 5.3, Table 6.2)
Accepted standard or way of behaving or doing things that most people agree with^

Normality, normalcy (Sects. 5.1, 5.4, 7.1)
"Situation where the basic functions of the society operate as usual and no particular perturbation affects its members" (Prêtre and Lochard 1995, 23)

Nuclear fuel cycle (Sects. 4.4.1, 5.2, 7.5.6)
Closed mining, processing, conversion, enrichment of uranium, nuclear fuel fabrication, reactor operation, electrical generation or other energy products, reprocessing to recover fissile material, storage of reprocessed fissile material, disposal (for highly radioactive fission products) and final end states for all waste*
Open mining, processing, conversion, enrichment of uranium, nuclear fuel fabrication, reactor operation, electrical generation or other energy products, storage of spent fuel, disposal and final end states for all waste*

Organisation (Box 5.1, 5.3, 6.1, Sects. 6.1, 7.3)
Group of people who work together in an organised way for a shared purpose^

Outreach (Sect. 2.7)
Effort to bring services or information to people where they live or spend time^ (esp. American English)

Oversight (Sects. 5.1, 5.6)
"implementing forms of control, which can result in immediate corrective actions, with or without regulatory force, and also as implementing forms of memory keeping …. Oversight is connected to the concept of continued societal responsibility" (NEA 2014b)

Ownership (Sects. 5.6, 6.2.2, 6.2.4, 7.1, 7.5.4, 7.7, Fig. 3.2, 6.4)

"desirable situation in which a community is not, and does not feel, dispossessed of plans and implementation … signifies that a community is empowered to define both problems and their solution (in appropriate partnership with other responsible actors)" (NEA 2013, 27)

Oxyfuel (Sect. 2.3)
Capture of CO_2 in power plants can be executed along three processes: post-combustion, pre-combustion or oxyfuel. In oxy-combustion, the fossil fuel is burnt with almost pure oxygen instead of air

Path dependency, path dependence (Sects. 5.5, 6.3, Table 6.2, Question 2 Chap. 2)
Decisions depend on prior decisions or past activities, foreclosing alternative ways

Pandemic (Sects. 5.4, 7.1, 7.4, Table 5.1, Fig. 7.3b)
(Of a disease) existing in almost all of an area or in almost all of a group of people, animals or plants^

Paradigm, scientific (enlarged: Sects. 3.2, 4.5.3, 4.7, 6.2.2, Table 4.3, Fig. 6.4)
Cf. Section 3.2. Constellation of beliefs, values, techniques, etc., shared by the members of a given [scientific] community (Kuhn 1996)

Participation (Sects. 5.7, 6.2.4, 7.5.2, Figs. 5.5, 6.4, 7.2, Table 6.2)
Cf. Figure 7.2. The fact that you take part or become involved in something^

Partitioning & transmutation, P&T (Sects. 5.2, 7.5.6, Table 4.1)
Partitioning is the selective separation of radiotoxic isotopes from reprocessing streams. After the successive partitioning has been done, the long-lived radionuclides are converted into shorter-lived or stable nuclides by a process called transmutation

Passive, passive component (Sects. 3.2, 3.3, 4.5.1, 5.2, Fig. 4.8, Table 3.1)
Component whose functioning does not depend on an external input such as actuation, mechanical movement or supply of power*

Peer review (Sects. 4.6, 7.7, Fig. 4.8)
Examination or review of commercial, professional or academic efficiency, competence, etc., by others in the same occupation*

Performance (Sects. 2.3, 4.6, Fig. 4.8, 5.3, Table 6.2)
How well a person, machine, etc., does a piece of work or an activity^

Performance assessment (Sects. 4.3, 4.4.1, 4.6, Table 4.5)
Safety analysis in the context of radioactive management
($\rightarrow$ risk assessment)

Perspective (Sects. 2.1, 2.6, 2.8, 3.1, 3.2, 3.3)
Particular way of viewing things that depends on one's
experience and personality^, a particular way of considering
something^

Performance assessment (Sects. 4.3, 4.4.1, 4.6, Table 4.5)
Plan (Box 6.1)
Set of decisions about how to do something in the future^

Planning (Sects. 6.1.2, 6.5, Fig. 6.2, Table 6.2)
Thinking about and deciding what you are going to do or
how you are going to do something^

Player (Box 7.1)
Here synonymous with Actor, agent, interested party,
stakeholder
Person, company, etc., with a concern or interest in the
activities and performance of an organisation, business,
system, etc.*

Pluralism, pluralistic (Sects. 3.4, 5.1, 5.2, 5.6, Fig. 6.4,
Box 7.1)
The existence of different types of people, who have dif-
ferent beliefs and opinions, within the same society^

Policy (Box 6.1, Sects. 6.1, 6.4)
Cf. Box 6.1. Set of ideas or plan of what to do in a particular
situation that has been agreed to officially by a group of
people, a business organisation, a government or a political
party^

Polluter-pays principle (Sects. 2.5, 4.4.1, Table 4.1, 4.4)
The idea that the person or organisation that causes pollution
should pay to put right the damage that it causes^

Precautionary principle
Taking action now to avoid possible environmental damage
when the scientific evidence for acting is inconclusive but
the potential damage could be great°
Approach/principle according to which the absence of full
scientific certainty shall not be used as a reason for postponing action where there is a risk of serious or irreversible
harm to the environment or human health. The
approach/principle is embedded in several instruments,
including Principle 15 of the 1992 Rio Declaration on
Environment and Development. Whereby the precautionary
approach is often used in negotiations to infer a less definite
meaning than the precautionary principle (UNEP, 2007)**
Where there are threats of serious or irreversible damage,
lack of full scientific certainty shall not be used as a reason
for postponing cost-effective measures to prevent environmental degradation (Rio Declaration 1992, principle 15)

"The precautionary principle serves the handling of risks
in situations where no acute danger is given. It aims at also
minimising risks which possibly manifest themselves only in
the long term and at preserving space for future developments" (Hilty and Som 2003, 29; transl. tf)

Pre-, post-combustion (Sect. 2.3)
Capture of CO_2 in power plants can be executed along three
processes: post-combustion, pre-combustion or oxyfuel.
Post-combustion uses a solvent to capture the CO_2 from the
flue gas, which requires low levels of nitrogen and sulphur
oxides, and therefore, requires energy-intensive de-NOx and
desulphurisation facilities. Pre-combustion involves reacting
a fuel with air or oxygen that contains CO and H_2. CO_2 is
separated and H_2 is used as a fuel

Problem (Boxes 1.1, 1.3, Figs. 1.1, 6.1, 6.2, 6.4, Sects. 3.2,
3.3, 5.5, 6.1.1, 7.5)
Cf. Box 1.1, Fig. 1.1, Sects. 6.1.1, 6.1.2. Situation, person or
thing that needs attention and needs to be dealt with or solved^
**Complex, environmental, ill-defined, ill-structured,
messy, wicked** cf. Box 1.1 and Fig. 1.1
Problem, implicit (Question 1 Chap. 1)
Implicit in this context means that a problem was caused,
superimposed, by a previous activity. It now constitutes a
factual constraint

Problem recognition (Sects. 3.3, 3.4, 5.6, 6.1.2, 6.2.1, 7.7,
Figs. 6.2, 6.5, Table 6.1)
Agreement that something is true or legal^ (here: that a
problem exists)

Procedure (Sects. 3.4, 4.6, 4.7, 5.6, 6.1.2, 6.2.3, 6.2.1, 6.2.3,
Box 5.3)
Series of specified actions conducted in a certain order or
manner.*
Set of actions that are the official or accepted way of doing
something^

Process (Sects. 2.8, 3.2, 3.4, 5.6, 6.1.1, 6.1.2, 6.2.1, 6.2.3, 6.
2.4, 7.2, 7.6, 7.7, Figs. 5.7, 6.2, 6.5, 7.1, 7.3, Table 4.5,
Boxes 5.1, 5.2, 5.3, 6.3)
Course of action or proceeding, especially a series of progressive stages in the manufacture of a product or some
other operation*
Set of interrelated or interacting activities that transform
inputs into outputs*
Series of actions that you take in order to achieve a result^

Programme (Sects. 3.4, 5.3, 5.5, 5.6, 5.7, 6.2.3, 6.5, 7.4,
7.5.5, 7.7)
Plan of activities to be done or things to be achieved^
Set of related measures or activities with a particular long-
term aim (https://www.lexico.com/definition/programme)
Planned series of future events or performances (ibid.)

Programme management

Process of managing several related projects, often with the intention of improving an organisation's performance°°°

Project (Sects. 1.2, 5.1, 5.6, 7.7, Fig. 3.2)

Piece of planned work or an activity that is finished over a period of time and intended to achieve a particular purpose^ Projects are undertakings that are time-bound (have an end), produce a unique output/deliverable and are executed under agreed-upon constraints (i.e., scope, timeline, budget/ resources) (https://blog.capterra.com/what-is-a-project/#1)

Protection (Sects. 3.1, 4.4.1, 4.5, 4.6, 6.2.1, 7.1, Fig. 3.1, Tables 3.1, 4.2)

Act of protecting or state of being protected^

Public (Sects. 2.7, 4.7, 5.1, 5.6, 6.2, 7.3, 7.6, 7.7, Figs. 3.2, 5.1, 5.2, Box 7.1, 7.4)

All ordinary people^ (UK, US), (US) all the people, esp. all those in one place or country. The general public (UK, US): ordinary people, esp. all the people who are not members of a particular organisation or who do not have any special type of knowledge^ Population of individuals in association with civic affairs, or affairs of office or state°°°

Publics (Sects. 6.2.3, 7.1, 7.4, 7.6, Box 7.1)

Small groups of people who follow one or more particular issues very closely. They are well informed about the issue(s) and also have a very strong opinion on it/them. They tend to know more about politics than the average person, and therefore, exert more influence, because these people care so deeply about their cause(s) that they donate a lot of time and money°°° (after Squire et al. 1995)

Purpose (Sects. 5.2, 6.1.1, 6.2.4, 7.5.5, Box 5.3)

Why you do something or why something exists^

Quality assurance (Sects. 4.3, 4.6, 5.1, Figs. 4.8, 7.1, Table 3.1)!

The function of a management system that provides confidence that specified requirements will be fulfilled*

Radioactive (Box 4.1)

Exhibiting radioactivity; emitting or relating to the emission of ionising radiation or particles*, here: as "opposed" to conventional (non-radioactive)

Radioactive material

Material designated in national law or by a regulatory body as being subject to regulatory control because of its radioactivity ("regulatory" meaning of radioactive)*

Radioactive waste (Figs. 3.2, 4.4, Tables 4.1, 4.2, 4.3, 4.4, 4.5, Box 4.1)

For legal and regulatory purposes, material for which no further use is foreseen that contains, or is contaminated with, radionuclides at activity concentrations greater than clearance levels as established by the regulatory body*

Radioactive waste management (Sect. 1.1, Figs. 3.2, 4.4, 6.4)

All administrative and operational activities involved in the handling, pretreatment, treatment, conditioning, transport, storage and disposal of radioactive waste*

Radionuclides (Sects. 4.5.3, 5.2, 7.5.6, Box 4.1, Table 4.2)

A radionuclide (radioactive nuclide or radioactive isotope) is an atom that has excess nuclear energy, making it unstable. This excess energy can be used in one of three ways: emitted from the nucleus as gamma radiation; transferred to one of its electrons to release it as a conversion electron; or used to create and emit a new particle (alpha particle or beta particle) from the nucleus. During those processes, the radionuclide is said to undergo radioactive decay°°° Primordial radionuclides ^{40}K, ^{235}U, ^{238}U, ^{232}Th and their radioactive decay products*

Rationale (Sects. 3.3, 3.4, 5.2)

The reasons or intentions that cause a particular set of beliefs or actions^

Rationality (Fig. 3.5)

The quality of being based on clear thought and reason, or of making decisions based on clear thought and reason^ **Bounded** rationality is limited, when individuals make decisions, by the tractability of the decision problem, the cognitive limitations of the mind and the time available to make the decision. Decision makers, in this view, act as satisficers, seeking a satisfactory solution rather than an optimal one. Therefore, humans do not undertake a full cost-benefit analysis to determine the optimal decision, rather they choose an option that fulfils their adequacy criterion°°° **Social** form of bounded rationality applied to social contexts, where individuals make choices and predictions under uncertainty, esp. situations in which not all alternatives, consequences and event probabilities can be foreseen°°°

Recycling (Sects. 4.4.1, 5.6, 7.5.6, Fig. 5.8, Table 4.1)

The (re-)processing of waste either into the same product or a different one°

Reflexive, reflexivity (Sects. 5.5, 6.2, 6.3, 7.2, 7.3, Box 1.3) Fact of someone being able to examine his or her own feelings, reactions and motives (= reasons for acting) and how these influence what he or she does or thinks in a situation^

Regulatory (body, etc.) (Sects. 2.4, 2.6, 3.4, 4.4.1, 5.8, 6.2.2, 6.2.3, 6.3, 6.4, 7.3, 7.7, Figs. 3.2, 5.7, Box 4.1, Tables 4.1, 4.4, 6.2)

Authority or system of authorities designated by the government of a state as having legal authority for conducting the regulatory process, including issuing authorisations, and thereby regulating the nuclear, radiation, radioactive waste and transport safety*

Reprocessing (Sects. 4.4.1, 5.6, 7.5.6, Box 4.1, Table 4.1) Process or operation, the purpose of which is to extract radioactive isotopes from spent fuel for further use*

Repository (Sects. 4.4.1, 4.5.1, 4.5.2, 5.1, 5.6, Figs. 3.4, 5.2, 6.5, Table 4.4, Box 7.2)
Engineered facility where waste is emplaced for disposal*

Research and development, R&D (Sects. 2.2, 2.5, 2.6, 7.1, 7.5.6, Fig. 2.1, Tables 4.1, 4.4)
Research, development and demonstration, RD&D (Sect. 4.5.3)
The part of a business/institution that tries to find ways to improve existing products/processes, and to develop new ones^. RD&D goes a step further than R&D in implementation

Residual waste (Sects. 4.4.1, 4.7, Box 4.1)
Waste remaining from materials after re-use, recycling and composting°

Resilience (Box 5.2, Sects. 5.3, 6.2.1, 7.5.4, 7.7, Question 1 Chap. 5)
Cf. Box 5.2. Ability of a substance to return to its usual shape after being bent, stretched or pressed^
Ability to be happy, successful, etc., again after something difficult or bad has happened^
Ecological capacity with which a system may absorb disturbance before having to be restructured (Gunderson et al. 2002)
Organisational human resilience is developed, sustained and grown through discourse, interaction and material considerations (Buzzanell 2010)

Resource (Sects. 2.3, 2.4, 4.5.1, 4.7, 5.4, 5.6, 5.7, 5.8, 6.1.3, 6.2.1, 6.2.2, 7.5.5, 7.5.6, Boxes 4.1, 5.2, 3.2, Figs. 3.4, 3.5, 7.1, Tables 3.1, 4.1, 4.4, 4.5, 4.6, 6.2, 7.1, 7.2, 7.7, Box 6.1, Questions Chap. 4)
Source or supply from which a benefit is produced and that has some utility°°°

Natural anything obtained from the environment to satisfy human needs and wants, from a broader ecological perspective, a resource satisfies the needs of a living organism°°°
Primary resource from nature (usable in its original form), such as freshwater or (via processing) ores, petroleums, wood (Question 2 Chap. 4)
Secondary resource to be recovered from human-made products or processes (Sects. 5.6, 7.5.6), Question 2 Chap. 4)
Common- or social-pool type of (congested) good consisting of a natural or human-made resource system (e.g., an irrigation system or fishing grounds), whose size or characteristics makes it costly, but not impossible, to exclude potential beneficiaries from obtaining benefits from its use°°° (Sect. 7.7)

Responsibility (Boxes 3.2, 5.3, Sects. 5.2, 5.6, 6.2.1, 6.2.2, 6.2.3, 7.1, 7.2, 7.3, Fig. 5.2, Table 6.2)
Moral status of morally deserving praise, blame, reward or punishment for an act or omission performed or neglected in accordance with one's moral obligations. Deciding what (if anything) counts as "morally obligatory" is a principal concern of ethics°°° (Sect. 7.2)
Social ethical framework suggesting that an individual has an obligation to work and cooperate with other individuals and organisations for the benefit of society at large°°° (Sects. 7.1, 7.2, 7.3)

Retrievability (Sects. 3.4, 4.4.1, 5.6, 6.2.1, Figs. 5.1, 5.7)
"ability in principle to recover waste or entire waste packages once they have been emplaced in a repository; retrieval is the concrete action of removal of the waste" (NEA 2013, 31)

Retrieval (Sects. 4.4.1, 4.5.1, 5.2)
Process of finding and bringing back something^
Effective action of recovering waste (→ Disposal, → Storage)

Reversibility (Sects. 3.4, 4.4.1, 5.6, 6.2.1, Figs. 5.1, 5.7)
If something is reversible, it can be changed back to what it was before^
"describes the ability in principle to reverse decisions taken during the progressive implementation of a disposal system; reversal is the actual action of going back on (changing) a previous decision, either by changing direction, or perhaps even by restoring the situation that existed prior to that decision" (NEA 2013, 35)
"possibility of reversing one or a series of steps in repository planning or development at any stage of the programme" (ibid.)
"implies a willingness to question previous decisions and a culture that encourages such a questioning attitude. It also implies some degree of retrievability of waste" (ibid.)

Review, esp. management system review (Sects. 3.4, 4.3, 4.4.1, 5.6, 6.1.2, Table 3.1, 6.2, Fig. 4.8, 5.1, 5.2, 6.2) (Regular and systematic) evaluation of an organisation of the suitability, adequacy, effectiveness and efficiency of its management system in executing the policies and achieving the goals and objectives of the organisation*

Peer review process of subjecting a scholarly work, research or ideas to the scrutiny of others who are experts in the same field, before a paper describing this work is published in a journal, conference proceedings or as a book°°° (Sects. 4.6, 7.7, Fig. 4.8)

Risk (Table 4.2, 4.3, 4.5, Figs. 3.5, 4.8, 4.2, Sects. 4.4, 5.4, 6.1, 7.2, 7.5.3, 7.7)

The possibility of something bad happening^

Detrimental health effects of exposure to a source or activity (including the likelihood of such effects occurring), and any other safety-related risks (including those to the environment) that might arise as a direct consequence of exposure, the presence of hazardous material (including waste) or its release to the environment; a loss of control over that material*

A multiattribute quantity expressing hazard, danger or chance of harmful or injurious consequences associated with exposures or potential exposures*

The mathematical mean (expectation value) of an appropriate measure of a specified (usually unwelcome) consequence*

Risk analysis … assessment … perception … acceptance … (Figs. 3.5, 6.4, Sects. 2.1, Chap. 4)

Risk management consists of risk identification, risk assessment (analytical), risk appraisal (normative, evaluation of options), risk decision and risk treatment/response (Flüeler 2006b). Obviously influencing risk appraisal, risk perception depends on the risk notion (risk definition, analysis, target), the hazard potential and the social context (Sjöberg and Drottz-Sjöberg 1994)

Robustness (Fig. 4.8, 5.2, 5.3, Sects. 3.4, 4.6, 5.1, 5.3, 5.8, Question 1 Chap. 5)

A system is robust if it is not sensitive to significant parameter changes, such as from external impact, and if it rests within well-defined boundaries (Weinmann 1991)

Robustness, social (Sects. 3.4, 5.1, 5.6, 6.2.3, 7.5.2, 7.7)

A decision is socially robust if most arguments, evidence, social alignments, interests and cultural values lead to a consistent option (Rip 1987)

Rolling present (Box 3.2, Sects. 5.3, 6.2.3, 7.5.4)

Normative principle to link present to future generations

Safety (Sects. 2.3, 4.5, 5.6, Tables 4.3, 4.5, Figs. 5.2, 5.3)

State in which or a place where you are safe and not in danger or at risk^

Protection of people and the environment against (radiation) risks, and the safety of facilities and activities that give rise to radiation risks*

Control of recognised hazards in order to achieve an acceptable level of risk°°°

"The concept of safety has evolved greatly within and outside the [NEA Forum on Stakeholder Confidence] community. Safety was firstly regarded as a technical and numerical concept …. The FSC community regards safety as more than just complying with technical requirements. Safety appears also to be tightly linked with societal and ethical concerns regarding decision-making processes and their outcomes (in terms of protection, fairness, etc.). Therefore, safety is not only a physical criterion, but also a social construct. As such, it has an emotional component as well. Technical safety and "peace of mind" safety are both goals. The "feeling" of safety inspired (or not) by a set of technical arrangements is a legitimate criterion, among many others, for judging those arrangements" (NEA 2013, 39)

Safety analysis (Sect. 4.5)

Usually called "performance assessment" in the context of radioactive waste, includes scenario development, conceptual and mathematical models development, consequence analysis, uncertainty and sensitivity analysis and confidence building (→ performance assessment)

Safety case (Sect. 4.5, Tables 4.3, 4.5)

Collection of arguments and evidence in support of the safety of a facility or activity, should acknowledge the existence of any unresolved issues and should provide guidance for work to resolve these issues in future development stages*

Safety culture (Sects. 6.4, 7.2, 7.3, Table 6.2)

Assembly of characteristics and attitudes in organisations and individuals which establishes that, as an overriding priority, protection and safety issues receive the attention warranted by their significance*

Safety function (Sect. 4.5.1, Tables 4.2, 4.5)

Specific purpose that must be accomplished for the safety of a facility or activity to prevent or mitigate radiological/harmful consequences of normal operation, anticipated operational occurrences and accident conditions*

Safety indicator (Fig. 4.8, Sect. 4.6)

A quantity used in assessments as a measure of the radiological impact of a source or of a facility or activity, or of the performance of protection and safety provisions, other than a prediction of dose or risk*

Scenario (Fig. 4.8, Sects. 2.3, 4.5.1, 4.5.3)

Description of possible actions or events in the future^

Postulated or assumed set of conditions and/or events in an assessment to represent possible future conditions and/or events to be modelled, such as the possible future evolution of a disposal facility and its surroundings*

Science and Technology Studies, STS (Sect. 2.1)

Study of how society, politics and culture affect scientific research and technological innovation, and how these, in turn, affect society, politics and culture°°°

Sequestration (Sects. 2.3, Figs. 2.2, 2.4)

Act of separating and storing a harmful substance such as carbon dioxide in a way that keeps it safe^ (→ carbon sequestration, → CCS)

Sewage sludge (Sect. 4.4.1)

By-product of wastewater treatment plants containing water and components removed from wastewater such as organic matter, nutrients and possible pathogens and heavy metals°°

Sievert (Tables 4.2, 4.5)

The SI unit of equivalent dose and effective dose, equal to 1 J/kilogramme (Sv, Millisievert, mSv)*

Sink (Sect. 4.4.1)

Material that cannot be recycled or recovered needs to go to the final sinks. They can be man-made (e. g., landfills) or natural (lakes, ocean). Systems that need after-care are merely temporary sinks, whereas final sinks pose no undue harm to man and the environment (Kral et al. 2019)

Siting (Sects. 4.6, 5.6, 5.7, 7.1, Figs. 3.2, 5.4, 7.1, Table 3.1)

Process of selecting a suitable site for a facility, including appropriate assessment and definition of the related design bases (including the concept and planning, area survey, site characterisation and site confirmation*

"process of identifying and developing a site for a waste management facility" (NEA 2013, 43)

Skill (Sects. 5.4, 7.2, 7.4, Box 3.2, 7.3, 7.4)

Ability to do an activity or job well, especially because you have practised it^

Ability to perform an action with determined results often within a given amount of time, energy or both°°°

Soft skills combination of interpersonal people skills, social skills, communication skills, character traits, attitudes, career attributes and emotional intelligence quotient (EQ) among others°°°

Society (Sects. 3.3, 5.4, 5.5, 6.2.2, 6.2.4, 7.3, 7.7, Figs. 3.1, 5.7, Box 5.1, 6.2, 6.3)

Large group of people who live together in an organised way, making decisions about how to do things and sharing the work that needs to be done. All the people in a country, or in several similar countries, can be referred to as a society^

Sociotechnical system (Sects. 1.3, 2.1, Fig. 1.5, Box 1.4)

Approach to complex organisational work design that recognises the interaction between people and technology in workplaces. Here: interaction between society's complex (waste) infrastructures and human behaviour°°°

Solution (Sects. 1.1, 2.3, 3.3, 3.4, 5.6, 6.2.1, 6.2.2, 7.5.4, Figs. 1.1, 6.2, 7.2, 7.7, Box 7.1, Question 1 Chap. 1, Question 3 Chap. 2)

The answer to a problem^

Spent (nuclear) fuel, S(N)F (Sects. 4.5.1, 5.1, 5.2, Box 4.1, Figs. 4.4, 4.5, 6.3)

Nuclear fuel removed from a reactor following irradiation that is no longer usable in its present form because of depletion of fissile material, poison buildup or radiation damage*

Stakeholder (Sects. 3.2, 3.3, 3.4, 6.2.2, 6.2.4, 7.1, Box 1.3, 7.1, 7.6, Tables 3.1, 6.2, Figs. 5.2, 5.7, 6.4, 6.5) (here also: actor, agent, player)

Person who acts for or represents another^

To "have a stake in" something, figuratively, means to have something to gain or lose by, or to have an interest in, the turn of events*

Stewardship (Sects. 5.3, 5.6, 6.2.3, Fig. 6.4)

Someone's stewardship of something is the way in which that person controls or organises it^

Storage (Sects. 2.3, 4.5.3, 5.2, 5.6, 6.2.1, Fig. 5.2, 5.11c, 6.3a, 6.5, Box 3.2)

Putting and keeping of things in a special place for use in the future^

Holding of radioactive sources, radioactive material, spent fuel or radioactive waste in a facility that provides for their/its containment, with the intention of retrieval*

Storage is, by definition, an interim measure, and the term [interim storage] would, therefore, be appropriate only to refer to short term temporary storage when contrasting this with the longer term fate of the waste*

Strategy, strategic (Boxes 6.1, 5.3, Sects. 2.3, 3.3, 4.4.1, 4.5.1, 5.5, 5.6, 6.1.2, 6.1.3, 6.2.1, 6.5, 7.1, 7.5.1, 7.7, Fig. 5.7, 6.2, Tables 6.1, 6.2)

Detailed plan for achieving success in situations such as war, politics, business, industry, or sport, or the skill of planning for such situations^

"Maintaining a balance between ends, ways and means; about identifying objectives; and about the resources and methods available for meeting such objectives … strategy comes into play where there is actual or potential conflict, when interests collide and forms of resolution are required" (Freedman 2013)

Structure, structured (Boxes 5.1, 1.1, 5.2, 7.5, Sects. 1.1, 6.2.1, 7.1, 7.5.2, 7.6, 7.7, Fig. 6.2, Table 6.2)!
The way in which the parts of a system or object are arranged or organised, or a system arranged in this way^
A general term encompassing all of the elements (items) of a facility or activity that contribute to protection and safety, except human factors.*

Surveillance (Sects. 5.1, 6.1.3, 6.4, 7.7, Table 4.3)
The careful watching of a person or place, especially by the police or army, because of a crime that has happened or is expected^
Type of inspection to verify the integrity of a facility or structure (here: not just physical inspection of a facility)*
Monitoring of behaviour, activities or information for the purpose of information gathering, influencing, managing or directing°°°

Sustainability (Figs. 3.1, 3.5, Sects. 3.2, 4.4.1, 6.2.1, 7.3, 7.5.5, Table 3.1, Box 7.6, Question 2 Chap. 3)
The quality of being able to continue over a period of time^
Ability to exist constantly°°°
Capacity for the biosphere and human civilisation to co-exist°°°

Sustainable development (Box 3.1, Sects. 2.1, 3.2, 3.4, 7.2)
Development that "meets the needs of the present without compromising the ability of future generations to meet their own needs" (WCED, World Commission on Environment and Development 1987)

Sustainable Development Goals (Sect. 2.3)
Collection of 17 interlinked goals set in 2015 by the United Nations to be a "blueprint to achieve a better and more sustainable future for all" to be achieved by 2030 (https://sdgs.un.org/)

Sustainable Development Scenario (Sect. 4.5.3)
Energy scenarios defined by the International Energy Agency to reach the climate goal pledged under the Paris Agreement of 2015: "holding the increase in the global average temperature to well below 2 °C above pre-industrial levels and pursuing efforts to limit the temperature increase to 1.5 °C above pre-industrial levels"
(https://www.iea.org/reports/sdg7-data-and-projections)

System (Boxes 5.1, 1.3, 5.2, 5.3, Sect. 1.3, Chap. 2, Sects. 3.1, 3.4, Chaps. 4, 5, Sects. 6.1.3, 6.2.1, 6.2.3, 7.2, 7.3, 7.5.4, 7.7, Figs. 1.5, 2.1, 3.1, 3.4, 4.2, 5.3, 5.7, 6.2, Tables 4.1, 4.2, 4.3, 4.4, 6.2)
Set of connected things or devices that operate together^

Systematically relevant (Sect. 5.4)
Institutions without which a system might be seriously destabilised (too important to fail), cf. critical infrastructures essential for the functioning of a society and economy°°°

System boundaries (Sect. 2.3)
Delimit which processes should be included in the analysis of a system, including whether the system produces any co-products that must be accounted for by system expansion or allocation (from life-cycle assessment)°°°

Target (Sects. 2.4, 2.5, 3.1, 4.4.1, 6.1.1, 6.2.1, 6.2.2, Figs. 1.1, 2.1, 4.8, 6.1, 6.2, Table 6.2)
Level or situation that you intend to achieve^

Technocentrism, technocentric (Table 3.1)
Value system that is centred on technology and its ability to control and protect the environment; environmental problems are seen as problems to be solved using rational, scientific and technological means°°°

Technology (Sects. 2.1, 2.5, 4.5.3, 5.1, 5.2, 7.3, 7.5.6, Fig. 2.1, 3.5, Boxes 1.3, 2.1, 3.2, 7.4)
(The study and knowledge of) the practical, especially industrial, use of scientific discoveries^
Science or knowledge applied to a definite purpose (Banta 2009)

Technology assessment (Box 2.1, Sects. 2.1, 5.5)
Scientific, interactive and communicative process that aims to contribute to the formation of public and political opinion on societal aspects of science and technology°°°, "form of policy research that examines short- and long-term consequences of the application of technology" (Banta 2009)

Thought structures, thought collectives (Box 7.5, Sects. 7.5.2)
Read Box 7.5

Time (Sects. 4.3, 5.2, Figs. 3.1, 3.2, 4.2, 5.6, Question 2 Chap. 5, Box 6.3, 7.2)
The part of existence that is measured in minutes, days, years, etc., or this process considered as a whole^

Tolerability, tolerable (Sect. 5.1)
Of a quality that is acceptable, although certainly not good^

Tolerance (Sect. 7.1)
Willingness to accept behaviour and beliefs that are different from your own, although you might not agree with or approve of them^

Toxic (Box 1.2, 4.1, Fig. 4.2, Sect. 2.3, Question 1 Chap. 1)
Poisonous^

Toxicity (Tables 4.2, 4.5)
The level of poison contained in a drug, or the ability of a drug to poison the body^

Toxin (Sect. 4.4.1)
Poisonous substance, especially one produced by bacteria, that causes disease^

Transdisciplinary (Box 1.3, 7.6, Sect. 7.7, → Disciplinary)
Research making "the change from research for society to research with society" (Scholz 2000, 13)

Transformation
Complete change in the appearance or character of something or someone, especially so that that thing or person is improved^
Transition of a sustainable path of Earth with, i.a., a post-fossil economic strategy (WBGU 2009)

Transgenerational (Sect. 7.7)
Transcending (many) generations

Transition (Sects. 2.3, 7.5.6)
Change from one form or type to another, or the process by which this happens^

Transpolitical (Sects. 5.2, 7.7)
Transcending day-to-day (partisan) politics (beyond terms of office)

Transscientific (Sects. 5.2, 7.7)
Questions "which cannot be answered by science" (Weinberg 1972)

Transmutation (Sects. 5.2, 7.5.6, Table 4.1)
→ Partitioning & transmutation

Transparency (Sects. 2.3, 5.6, 6.2.2, Table 6.2)
Characteristic of being easy to see through^
"attribute of a process seeking to reveal, in a non-adversarial manner, the information, values and assumptions present behind the arguments or activities of each type of stakeholder" (NEA 2013, 57)

Trust (Sects. 5.3, 5.4, 5.6, 6.1.2, 6.2.1, 6.2.3, 6.3, 7.3, 7.7, Fig. 5.2, Tables 6.1, 6.2)
Belief that someone is good and honest and will not harm you, or that something is safe and reliable^
"Related to the behaviour of individuals and organisations; it has to be earned, and it is related to feelings of comfort and liking … signifies that an individual is willing to give up a certain measure of control to another person, an institution, or a set of institutions" (NEA 2013, 9)

Uncertainty (Sects. 3.3, 4.5.1, 4.6, 5.5, 6.1.1, 6.2.1, 7.5.4, Fig. 4.8)
Aleatory uncertainty uncertainty inherent in a phenomenon. Aleatory uncertainty (or stochastic uncertainty) is taken into account by representing a phenomenon in terms of a probability distribution model*
Epistemic uncertainty uncertainty attributable to incomplete knowledge about a phenomenon, which affects the ability to model it, reflected in a range of viable models, multiple expert interpretations and statistical confidence*

Unintended consequences (Sects. 2.3, 5.5)
(Unplanned and unwanted) result of a particular action or situation, often one that is bad or not convenient^

Unknown unknowns (Sect. 2.3)
From "Johari" window (Joseph Luft and Harrington Ingham, cognitive psychologists) to better understand oneself and others: 4th quadrant ("not known to self" and "not known to others")°°°

Validation (Figs. 5.3, 6.2)
The process of determining whether a product or service is adequate to perform its intended function satisfactorily. Validation (typically of a system) concerns checking against the specification of requirements, whereas verification (typically of a design specification, a test specification or a test report) relates to the outcome of a process. Key question: Do we do the right thing?*

Verification (Sect. 2.4)
The process of determining whether the quality or performance of a product or service is as stated, as intended or as required. Key question: Do we do it in the right way?*

Voluntariness, voluntary (Sect. 5.6)
Done, made or given willingly, without being forced or paid to do it^

Voluntarism (Sect. 3.4)
"Local government representatives of a community express an interest in participating in a process to determine the suitability of siting a radioactive waste management facility within the boundaries of their community. The ability for partnership members to come to the table as equal partners is also important" (NEA 2013, 21)

Vulnerability, vulnerable (Sects. 2.3, 5.4, 6.2.2, Box 6.2)
Able to be easily physically, emotionally or mentally hurt, influenced or attacked^
Degree to which a community, population, species, ecosystem, region, agricultural system or some other quantity is susceptible to, or unable to cope with, adverse effects of climate change (UNEP, 2007)**

Waste (Box 4.1; 1.2, Fig. 4.1)
Material for which no further use is foreseen*
Substance or object which is disposed of or intended to be disposed of or required to be disposed of by the provisions of national law (Basel Convention)**
Secondary waste (radioactive waste) resulting as a byproduct from the processing of primary (radioactive) waste*
High-level waste (HLW) the radioactive liquid containing most of the fission products and actinides present in spent fuel—which forms the residue from the first solvent extraction cycle in reprocessing—and some of the associated waste streams; this material following solidification; spent

fuel (if it is declared as waste); or any other waste with similar radiological characteristics* (consult entries above) (Box 4.1, Tables 4.1, 4.2, Sects. 1.1, 4.4.1, 4.5.1, 5.1, 5.2, 5.3, 6.2.1, Figs. 4.4, 5.11, 7.1)

Intermediate-level waste (ILW) radioactive waste that, because of its content, in particular its content of long-lived radionuclides, requires a greater degree of containment and isolation than that provided by near-surface disposal* (consult entries above) (Box 4.1, Tables 4.1, 4.2, Sects. 4.5.1, 5.2, 5.6, Figs. 4.4, 7.1)

Long-lived waste radioactive waste that contains significant levels of radionuclides with a half-life greater than 30 years* (Box 4.1, Sect. 4.5.1, Fig. 4.8)

Low-level waste (LLW) radioactive waste that is above clearance levels, but with limited amounts of long-lived radionuclides* (Box 4.1, Sects. 2.3, 4.4.1, 4.5.1, 5.6, 6.2.4, Tables 4.1, 4.2, 4.5, Figs. 4.4, 7.1)

Short-lived waste radioactive waste that does not contain significant levels of radionuclides with a half-life greater than 30 years* (Box 4.1, Sects. 2.3, 5.2, Tables 4.2, 4.5, Fig. 4.4)

E-waste (Fig. 5.9, Sect. 7.5.6)

Popular, informal name for electronic products nearing the end of their "useful life." Computers, televisions, VCRs, stereos, copiers and fax machines are common electronic products. Many of these products can be reused, refurbished or recycled°°

Worldview (Table 3.1, Sect. 5.2)

Set of beliefs about fundamental aspects of reality that ground and influence all your perceiving, thinking, knowing, and doing. Fundamental cognitive orientation of an individual or society encompassing the whole of thfe individual's or society's knowledge and point of view°°°

Sources:

(all weblinks accessed 27 January 2023)

*International Atomic Energy Agency https://www.iaea.org (Glossary below)

**https://www.genevaenvironmentnetwork.org/resources/glossary/

***https://geology.com/dictionary/glossary-a.shtml

°https://www.planningportal.co.uk/

°°United Nations University, https://flores.unu.edu/en/research (glossary)

°°°https://en.wikipedia.org/wiki/Main_Page

^https://dictionary.cambridge.org/dictionary/english/

^^https://www.nagra.ch/en/types-of-radioactive-waste

Argyris C (1991) Teach smart people how to learn. Harvard Business Review 69(3):99–109; Reflections 4(2):4–15

Banta D (2009) What is technology assessment? International Journal of Technology Assessment in Health Care 25(Supplement S1)7–9. https://doi.org/10.1017/S0266462309090333

Drucker PF (1963) Managing business effectiveness. Harvard Business Review (3):53–60. https://hbr.org/1963/05/managing-for-business-effectiveness

Hilty L, Som C (2003) Vorsorgeprinzip, Nachhaltigkeit und ethische Aspecte der Informationsgesellschaft [The precautionary principle, sustainability, and ethical aspects of the information society]. In: Hilty L et al. (eds) Das Vorsorgeprinzip in der Informationsgesellschaft. Auswirkungen des Pervasive Computing auf Gesundheit und Umwelt. TA 46/2003. TA-SWISS, Bern, pp 29–45

IAEA (2019) IAEA safety glossary. Terminology used in nuclear safety and radiation protection 2018 edition. IAEA, Vienna. https://www.iaea.org/publications

Kral U, Morf LS, Vyzinkarova D, Brunner PH (2019) Cycles and sinks: two key elements of a circular economy. Journal of Material Cycles and Waste Management 21:1–9. https://doi.org/10.1007/s10163-018-0786-6

Kühl J (2017) Failure culture—the key ingredient for innovation leadership. Center for Digital Technology & Management, CDTM, Munich. https://medium.com/cdtm/failure-culture-the-key-ingredient-for-innovation-leadership-37735a48057a

NEA (2013) Stakeholder confidence in radioactive waste management. An annotated glossary of key terms. No. 6988. OECD, Paris. https://www.oecd-nea.org/upload/docs/application/pdf/2019-12/6988-fsc-glossary.pdf

NEA (2014) Control, oversight and related terms in the international guidance on geological disposal of radioactive waste—review of definitions and use. NEA/RWM/RF(2014)2. OECD, Paris. https://www.oecd-nea.org/jcms/pl_19460

Sjöberg L, Drottz-Sjöberg BM (1994) Risk perception. In: IAEA (ed) Radiation and society: comprehending radiation risk. IAEA Conference, 24–28 Oct 1994, Paris, vol 1: Comprehending radiation risks. A report to the IAEA. IAEA, Vienna, pp 29–59

Squire P, Lindsay JM, Covington CR, Smith ERAN [Blasco S] (1995/2017, multiple editions) Dynamics of democracy. Brown & Benchmark, Madison, WI/Atomic Dog Publishing, Mason, OH

WBGU (2011) World in transition—a social contract for sustainability. WBGU German Advisory Council on Global Change, Berlin. https://www.wbgu.de/en/publications/publication/world-in-transition-a-social-contract-for-sustainability#section-downloads

Other references are given at the end of each chapter.

Printed in the United States
by Baker & Taylor Publisher Services